Enhancing Entrepreneur

To Samantha and Melissa, with my heartfelt wishes that all the possibilities ***they*** *choose to follow will be happy ones.*

Enhancing Entrepreneurial Excellence

Tools for making the *possible* real

Robert A. Baron

Oklahoma State University, USA

Edward Elgar
Cheltenham, UK • Northampton, MA, USA

Published by
Edward Elgar Publishing Limited
The Lypiatts
15 Lansdown Road
Cheltenham
Glos GL50 2JA
UK

Edward Elgar Publishing, Inc.
William Pratt House
9 Dewey Court
Northampton
Massachusetts 01060
USA

A catalogue record for this book
is available from the British Library

Library of Congress Control Number: 2012948849

This book is available electronically in the ElgarOnline.com
Business Subject Collection, E-ISBN 978 1 78195 209 2

ISBN 978 1 78195 208 5 (cased)

Typeset by Servis Filmsetting Ltd, Stockport, Cheshire
Printed and bound by MPG Books Group, UK

Contents

About the author vi
Preface: Changing the world, one idea at a time vii
Acknowledgments ix

1 The nature of entrepreneurship—and why entrepreneurs truly matter 1

2 Motivation: why entrepreneurs make the journey 19

3 Human cognition: the internal origins of creativity, innovation, and ideas for the *possible* 38

4 From desire to achievement: the crucial role of self-regulation 62

5 The social side of entrepreneurship: getting the help you need 83

6 The personal side of entrepreneurial excellence: characteristics that enhance success 103

7 Making effective decisions—and avoiding cognitive traps 128

8 Managing adversity: dealing with stress, learning from our mistakes, and coping with failure 148

9 Putting it all together: a model of the highly effective entrepreneur 170

Author index 187
Subject index 190

About the author

Robert A. Baron (PhD, University of Iowa) is the Spears Professor of Entrepreneurship at Oklahoma State University. He has held faculty appointments at RPI, Purdue, the Universities of Minnesota, Texas, South Carolina, Washington, Princeton University, and Oxford University. He served as a Program Director at NSF (1979–1981), and was appointed as a Visiting Senior Research Fellow by the French Ministry of Research (2001–2002; Toulouse). He has served as a Department Chair and as Interim Dean of the Lally School (RPI). He holds three U.S. patents and was founder and CEO of IEP, Inc. Baron's current interests focus on the role of cognitive and social factors in entrepreneurship. He has published more than 135 articles, 36 chapters, and 45 books. Professor Baron serves on the boards of AMJ, JBV, and JOM, and is an Associate Editor for SEJ.

Preface: Changing the world, one idea at a time

Change is indeed the essence of life in the early twenty-first century, but even so, sometimes—when I pause to consider the breadth of the changes I've witnessed during my life—I am truly astonished. When I was a child in New York City (in the dim-and-distant 1950s), we often rode on trolley cars (what Europeans call trams); and they were found not only in large cities, but in every sizeable town, including the one in which I live now (Edmond, Oklahoma). Today, of course, trolleys exist only in memory, in museums, and in a small number of U.S. cities. Television was brand new, screens were tiny, showed only black and white images, and—even worse—there was almost nothing to watch! Frozen food was a new fad, and from it, I soon learned lasting lessons about the misleading nature of advertising; in my ten-year-old mind, the foods inside should, I thought, resemble pictures shown on the front of the packages. Of course they did not, and this gave me an early taste of media-based disillusionment! Few people I knew had ever flown on an airplane, and fewer still had ever visited Europe or even Mexico. Having a college degree was a relative rarity, and I knew very few women who worked full time outside their homes. Air conditioning was a luxury few could afford, and was to be found mainly in movie theaters and restaurants. As for technology, small communication devices existed only in comic books or science fiction stories, computers were huge devices used only by a few people who understood their deep mysteries; music came on black, breakable (and later, unbreakable) discs, tape recorders had actual reels that often got tangled, and no one had ever heard the term "download". The only way to capture moving images of other people was by means of 8 mm cameras, and the idea of shopping electronically would have seemed totally incomprehensible. And medical procedures—even relatively basic ones such as removal of an appendix—required long stays in the hospital (eight days, when my own appendix was removed in 1951); now, due to new surgical techniques, a one-day stay will often suffice—as was true, in recent years, for both my daughter Jessica and wife Rebecca! So, yes, the world—and more importantly, everyday life—has changed tremendously during my lifetime.

Why have all these transformations occurred in just a few decades? Obviously, many factors are involved, ranging from rapid progress in science and engineering, which provided the foundation for many advances, through rising standards of living and education, as well as the huge sums spent by large corporations and the military on research and development. But overall, I believe—passionately!—that a key driver of all this change has been the ideas created by individuals who then, through a long and complex process, converted them into real products, services, or means of production. In short, I believe that it is these individuals—the individual entrepreneurs—who, through their collective creativity and efforts, have given us the world in which we now live.

Of course, having an idea is one thing; developing it into something tangible is quite another. The process involved is long and complex, and to accomplish it, entrepreneurs need many skills and personal resources—tools that are helpful, if not truly essential, for making the *possible* real. Identifying these as clearly as current evidence permits is the key goal of this book; offering suggestions that can help entrepreneurs to develop them, is another. Each chapter, then, examines distinct tools useful to entrepreneurs in their journey from ideas to reality, and together, the information included offers as broad and comprehensive an overview as I could produce, of what these tools are and why they are important. Many entrepreneurs fail, and it is a central thesis of this volume that in many instances, they do so because they are lacking in some of the essential tools or resources.

This logic has led me, on many occasions, to ponder yet another question: "How many potentially valuable ideas 'die on the vine'—are generated by individuals (potential entrepreneurs), but are then never developed?" My musings suggest to me that this may be a much larger number than most of us would prefer to imagine. And these thoughts also lead me to ask, "How many such ideas are fading away, unknown, even today?" If this book, and the contents it presents, can help assure that even a few of these ideas do actually see the light of day (because the people who generate them have gained a better understanding of what they need to achieve this transformation), then it will have accomplished its major purpose.

In any case, get ready for an intriguing journey—one that will take us from the cognitive origins of ideas, through examination of the many factors that influence their development, and the likelihood that they will ever become more than ephemeral images of something new and better flickering through the minds of specific persons. It is far better, I believe, for these ideas to actually move from the realm of the possible into reality. Why? Because such transformations benefit not only the persons who create the ideas, but the rest of us too, by changing the world in ways we cannot now predict or, perhaps, even imagine. So, please permit me to close with a warm salute to entrepreneurs—who are, in a very real sense, the true engines of change and improvement in human life.

Acknowledgments

Some words of thanks

Writing itself is a solitary task, best carried out in a quiet (and sometimes lonely!) place, designed to minimize distractions. But despite this fact, many talented and dedicated people always contribute to the final product. Here, I'd like to thank some of these individuals.

First, my sincere thanks to my Editor, Alan Sturmer. His input helped shape the basic idea for this book and its contents, and his faith in my ability to produce (!) it has never wavered. So, "Thank you", Alan for your support and expert advice.

Next, I want to thank Alison Hornbeck for preparing the manuscript for production—a crucial step and one that must be done very carefully, which she certainly did!

Thanks are also due to Elizabeth Clack for overseeing and directing the production process—truly a key task, and to Sarah Cook for a very careful job of copy-editing, which for me, personally is especially crucial. I know only too well that I make many errors, and calling these to my attention so that they can be corrected is an essential aspect of the process.

Thanks too, to Jennifer Wilcox for preparing the catalog entry, and to Katy Wight for her help with the author information form—a short document, but one that includes important information.

Truly special words of thanks are due to Mr. Dale Chihuly, who granted me permission to feature his amazing, creative art on the cover of this book. I have always been deeply moved by Mr. Chihuly's creations, and cannot imagine a better way of illustrating how the human spirit (and a high degree of talent!) combine to give concrete form to the *possible*.

Last (but of course ***not*** least!), my warmest appreciation to my wife Rebecca, whose comments and suggestions throughout the months of putting my thoughts on paper, were—as always—creative, constructive, and truly invaluable.

Robert A. Baron
Stillwater, Oklahoma

1 The nature of entrepreneurship—and why entrepreneurs truly matter

Chapter outline

Entrepreneurship—new ventures . . . and beyond
Why entrepreneurs matter—and why they really *are* different
Are entrepreneurs different? And if so, *why*?
Acquiring knowledge about the foundations of entrepreneurial success
Asking entrepreneurs: can they really tell us the "secrets of success"?
Conducting systematic research
Importing knowledge from other fields
Tools for making the possible real: a brief overview

"It's always best to start at the beginning", the Good Witch tells Dorothy in the *Wizard of Oz*, and then she points to the yellow brick road, unrolling invitingly before them. "Start at the beginning" is good advice in almost any context, but what, precisely, *is* the right beginning for a book such as this? There is no yellow brick road to show the way, but like Dorothy, who wanted desperately to return to Kansas, we do have an ultimate goal. Basically, it is this: gaining an accurate, comprehensive understanding of the *tools* entrepreneurs need to achieve excellence—the tools they need to succeed in their basic task which is—precisely what the title of this book suggests: making the *possible* (their ideas, dreams, and inspiration) real.

With that central goal in mind, the key questions to address in this initial chapter come sharply into focus. First, it is important to define the nature and scope of entrepreneurship—to indicate clearly what it involves and does not involve. Does it consist entirely of starting new ventures (businesses)? Or can it take other forms as well?

Second, what is the role of entrepreneurs in this process? Are they truly, as this book suggests, at its very center? Or are they, perhaps, less important than we might at first believe? Your first reaction is probably "*Of course* entrepreneurs are central!" In fact, though, there is another view suggesting that individual entrepreneurs, and their personal characteristics or skills, are less important than external conditions and contexts. According to this view (which derives primarily from the field of economics) if an opportunity emerges as a result of technological advances, changes in society, or other factors, *someone* will recognize it and develop it, and who that person is matters little, if at all.

As the title of this book suggests, it strongly supports the first view—the idea that entrepreneurs, as individuals, do play a crucial role, and that trying to understand the entrepreneurial process without careful attention to them is, as one scholar (who happens to be an economist!) put it: "like trying to understand Shakespeare without including Hamlet" (Baumol, 1968). In other words, individual entrepreneurs are *not* interchangeable parts of a complex economic system or mechanism in which they play only a limited role; rather, their skills, knowledge, motives, values, personal characteristics, and actions do matter in the sense that they strongly shape both the process and its ultimate outcomes—which can range from tremendous success to total failure. Although this book consistently adopts the first "entrepreneurs do matter" view, the contributions and implications of the second "external conditions are central" perspective will also be noted. In addition, another basic question will also be briefly considered in this initial chapter: "Do entrepreneurs really differ from other persons—and if so, why?"

After discussing these issues, we will turn briefly to yet one more: How can knowledge about the foundations of entrepreneurial excellence be acquired? Simply by asking entrepreneurs for their views? By conducting systematic research on this question? By importing knowledge from other fields that have long studies the foundations of success? The position adopted here is that all three approaches can be useful and are, in fact, complementary rather than competing. As a result, information provided by all will be included in later chapters.

Finally, we will provide an overview of the various tools that will be examined in detail in later chapters. These are highly varied in nature, but together provide important insights into the basic foundations of entrepreneurial success.

Entrepreneurship: new ventures . . . and beyond

In one sense, we are living in what could be described as the "decade(s) of the entrepreneur". Entrepreneurship is featured in countless magazine and newspaper articles, popular television shows (e.g., *Shark Tank*) and major films (*The Social Network*). Universities have jumped on board by adding courses, departments, and even Schools of Entrepreneurship to their offerings; governments at all levels seem to be competing to develop programs and incentives to encourage entrepreneurship. A key reason behind this amazing popularity is the strong conviction that entrepreneurship is a powerful engine of economic growth. Little wonder, then, that interest in it has grown exponentially in recent years.

Despite this intense interest, however, there is very little public attention to the following question: *What, precisely, is entrepreneurship?* For most persons, this is a non-issue: entrepreneurship is, simply, what entrepreneurs do—and what they do is start new ventures and, if they are fortunate, become fabulously rich! The term "entrepreneur", then conjures images of people such as Bill Gates, Steve Jobs, Mark Zuckerberg, and more recently, Kevin Systrom and Mike Krieger, founders of Instagram, who recently sold their company to Facebook for $1 billion. But is this view of entrepreneurship accurate? Does it consist entirely of starting a new venture to develop an emergent business opportunity? Many courses and books on entrepreneurship assume that this is so. And certainly this activity is a key way in which entrepreneurs convert their ideas into reality—products, services, processes, and so on.

In fact, however, this is not the entire story. Rather, as Shane (2012), co-author (with S. Venkataraman, 2000) of perhaps the most influential definition of entrepreneurship to date, suggests that entrepreneurship is broader in scope than simply forming new ventures. Consistent with this view, therefore, entrepreneurship is defined, throughout this book as follows: *the application of human creativity, ingenuity, knowledge, skills, and energy to the development of something new, useful, and better than what currently exists—something that creates some kind of value (economic, social, or other)*. In other words, entrepreneurship occurs whenever, wherever, and however, individuals take concrete action to convert their ideas and dreams about "something better" into reality—into something tangible such as a new product or service (Baron, 2012). Reaching this goal certainly does often involve launching and developing a new business; that is often a central aspect of entrepreneurship and will be a key focus of the present book. But it is important to note that this is not the only way in which human creativity can be expressed in efforts to develop something that is new, useful, and better than what already exists. In fact, individuals can think and act entrepreneurially without necessarily launching a new venture. What is essential is simply that they *take action to move their ideas from the realm of the possible to concrete reality*.

Although the view of entrepreneurship described above does not align entirely with commonly held views of this activity, it is actually closely aligned with the well-known and widely accepted definition mentioned above (Shane and Venkataraman, 2000). Slightly paraphrased, this definition suggests that: *Entrepreneurship, as a field of business, seeks to understand how opportunities to create something new (e.g., new products or services, new markets, new production processes or raw materials, new ways of organizing existing technologies) arise and are discovered or created by specific persons, who then use various means to exploit or develop them, thus producing a wide range of effects.* In essence, this definition suggests that entrepreneurship does indeed involve the conversion of ideas (recognized or created opportunities) into something new and tangible through some kind of overt action. This can involve starting a new business, and often does. But entrepreneurship, this definition suggests, can also occur in many other contexts and take many other forms. For instance, consider a talented teacher who realizes that currently popular techniques for equipping students with basic skills are ineffective. The teacher, on the basis of her or his experience, and talent, develops a new approach—one that is, in fact, better than the present techniques. This teacher is thinking and acting entrepreneurially, even if she or he does not start a new company to develop and promote these new methods.

Similarly, consider a physician who realizes that many patients she treats are not following her advice—for instance, they are not taking their medicines regularly and on schedule. To improve this situation, she gives each patient an inexpensive electronic device programmed to deliver "reminder signals" at appropriate times. The device is very small, so patients can keep it with them all the time. To the extent this new approach changes her patients behavior so as to make her treatment of them more effective, the physician has acted entrepreneurially: she has applied her ingenuity and creativity to developing something new and better—a technique that is not now a standard part of medical practice, but which may help to improve her patient's health—and her effectiveness as a physician. Could the physician then start a company to promote and sell this device? Absolutely. But even if she does not, she has already acted entrepreneurially.

To recapitulate: a central tenet of this book is that entrepreneurship, as an activity, is actually broader in scope than the traditional view which focuses on starting new ventures. In fact, it occurs whenever, wherever, and however, human beings use their own creativity, ingenuity, energy, knowledge, and talent to develop something new, useful, and better. This will be a basic theme in the remainder of this volume, and although much attention will certainly be focused on new ventures (this is the source of much of the evidence we will include), this broader perspective will be reintroduced often as appropriate.

Why entrepreneurs matter—and why, perhaps, they *are* really different

It is a sad fact of life—a very sad fact!—that most new ventures do not survive or prosper. As shown in Figure 1.1, fewer than half are still

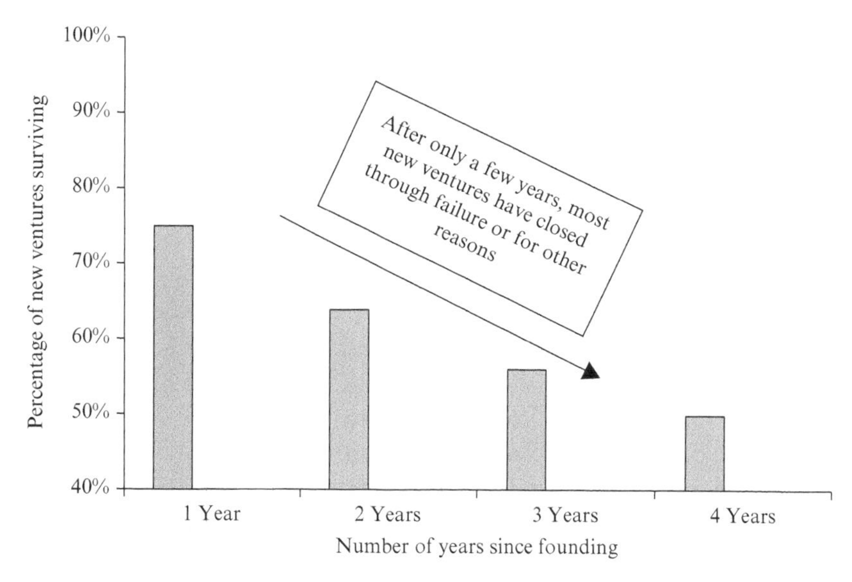

Note: As shown here, a large proportion of new ventures disappear within a few years. This does not necessarily mean they have failed, in a business sense, but in any case, they are no longer in existence and doing business.

Source: Based on data from Shane, 2009.

Figure 1.1 *New ventures failures: all too common*

operating four years after being founded and only 20 percent are in existence a few years later. This raises a very puzzling and fundamental question: *Why*? Clearly, these new businesses start with high hopes and enthusiasm on the part of their founders—in fact, founding entrepreneurs often make them the center of their lives. Why, then do so many disappear so quickly? By the way, it is important to note not all are failures: some cease to exist because their founders voluntarily decide to close them, often in order to sell the company or do something else, such as founding another new business! Here, though we will focus on business closings that are due to failure.

To return to the central question: Why do so many new ventures vanish so quickly? This question has been the subject of a great deal of attention, and countless articles about it have appeared in magazines, on-line forums, and books. An informal survey of many such sources indicates that among the factors mentioned most frequently as playing a key role in new venture failure are the following:

1. Starting the business for the "wrong" reason: for instance, too intent on getting rich very quickly.
2. Lack of a sizeable market for the new venture's product or services.

3. Undercapitalization: not enough funds to survive high start-up costs.
4. No obvious competitive advantage over existing products or services.
5. Expanding too quickly.
6. Competing directly with established industry leaders.
7. Lack of sufficient, high-quality planning.
8. Too much leverage—assuming too much and too much risk.
9. High overheads.
10. Poor location—far from customers, suppliers, and so on.

Together, these potential causes suggest that new ventures can, and often do, "go off the rails" for many reasons. Although these factors are varied in scope, careful scrutiny of them suggests that many involve *actions, judgments, miscalculations, or errors* by the founding entrepreneurs. Yes, external circumstances certainly play a role—for instance, existing competition is often intense, markets change rapidly, and technology advances at breathtaking speed. Yet, in essence, many new venture failures seem to be related to something that their founders *did not* do that they should have done (e.g., conduct careful market research; minimizing overheads; obtaining sufficient start-up funds), or something they *did* do that they should *not* have done (e.g., choosing to compete directly with entrenched industry leaders, seeking very rapid growth too soon, assuming too much debt). In short, many business failures derive, ultimately, from what might be termed, in another context (e.g., plane crashes, accidents at nuclear power plants) human error.

But why does this occur—why do entrepreneurs contribute, so often, to the failure of their own new ventures? The view adopted in this book is straightforward: often, entrepreneurs—motivated, talented, and dedicated though they are—are lacking in certain tools essential for achieving success. Perhaps a concrete example will help clarify this view; and nothing illustrates it more clearly than the following sad chain of events.

Who invented the telephone? Almost certainly your answer is "Alexander Graham Bell". He founded Bell Telephone and became tremendously famous and wealthy as "his" invention gained widespread use, so he is generally credited with originating this world-changing device. But was he the real inventor of the telephone? History indicates that in fact, the credit for developing a practical, working telephone should go to Antonio Meucci. Meucci knew Bell, had worked with him, and had developed a working telephone several years before Bell filed his initial patent application. When he sent a working prototype of his invention to Western Electric for testing, however, it "mysteriously" disappeared and could not be found when, after several months, Meucci asked that it be returned. If you have never heard of Antonio Meucci, however, do not be surprised: almost no one has. He totally lost the race to patent the telephone, and died in obscurity—and poverty. Bell, in contrast, proceeded at full speed with development of the same product, and soon obtained a patent for it. Did he steal Meucci's invention? That is a complex question, but consider this: In 1991, the United States Congress issued the following resolution (printed, in part below):

> Whereas in March 1876, Alexander Graham Bell, who conducted experiments in the same laboratory where Meucci's materials had been stored, was granted a patent and was thereafter credited with inventing the telephone;
> Whereas Meucci never learned English well enough to navigate the complex American business community;
> Whereas Meucci was unable to raise sufficient funds to pay his way through the patent application process, and thus had to settle for a caveat, one year renewable notice of an impending patent, which was first filed on December 28, 1871;
> Resolved, That it is the sense of the House of Representatives and the Senate that the life and achievements of Antonio Meucci should be recognized; and the work of Antonio Meucci in the invention of the telephone should be acknowledged.

Why did Bell, and not Meucci, gain the credit—and wealth—associated with this important invention? The answer is suggested, in part, by the Congressional resolution: Meucci was an immigrant from Italy who did not speak English well, and was therefore hampered in his efforts to persuade investors to support his work; further, he was shy and retiring by nature, and also suffered from ill health. In contrast, Bell was an outstanding and fluent speaker, was well-connected in the business world, and enjoyed excellent health and vigor throughout his life. One conclusion: Bell "won" because he possessed tools essential for achieving success—tools (skills, capacities, characteristics) that Meucci lacked (see Figure 1.2).

Certainly, this sad tale does not in itself "prove" that entrepreneurs matter, and that their success is determined, to an important degree, by their personal attributes. In fact, as noted earlier, a contrasting view, suggesting that highly successful people—including entrepreneurs—are successful not because of their personal characteristics, but largely because of external circumstances, has received considerable attention. Perhaps the most eloquent statement of this perspective is provided by a best-selling book by Malcolm Gladwell, *The Outliers* (2008). In this volume, Gladwell suggests—very eloquently—that when and where someone is born, and key aspects of her or his culture, are just as important, if not more important in determining success, than intelligence, drive, talent, and other personal characteristics. For instance, Gladwell notes that although Bill Gates was certainly a talented and highly focused young man (he spent countless hours in a nearby computer laboratory), he could not have developed the

Note: Although Antonio Meucci (left) was the actual inventor of the telephone, Alexander Graham Bell (right) patented this invention—and grew rich from it. One explanation for these outcomes is that Bell possessed skills and characteristics that Meucci lacked (e.g., he was a persuasive fluent speaker of English; had better connections in the business world, etc.).

Sources: Photograph Courtesy of the Garibaldi-Meucci Museum, Staten Island, NY (left); Library of Congress, Prints and Photographs Division, photograph by Harris & Ewing, [LC-H25- 11186-E] (right).

Figure 1.2 *The consequences of lacking essential tools*

ideas that lead him to found Microsoft (and become fabulously wealthy), if he had been born somewhere else (a location where access to computing equipment was unavailable), at the "wrong time" (i.e., before the knowledge needed to develop a practical modern personal computer existed), or to a less supportive and affluent family.

In short, this "context" or "external factors" view downplays the role of the kind of tools on which book will focus (i.e., the skills, motives, knowledge, characteristics, etc. of individuals). However, even Gladwell acknowledges that these characteristics play *some* role; he simply contends that they are less important than has often been believed. In essence, his position is that although not anyone could have founded Microsoft (Bill Gates' personal characteristics did indeed matter to some extent), many people, given the advantages he enjoyed, could.

When disagreements such as this occur in science, the best way to resolve them is through careful analysis of existing evidence plus the collection of additional data. And in this instance, doing so suggests the following conclusion: Yes, external factors do matter—for instance, no one could have developed a tablet computer (e.g., an iPad, Kindle, or Nexus) in the absence of certain kinds of supporting technology; and no one would devote time and energy to developing techniques for growing truffles (a kind of fabulously expensive mushroom) in a culture that had no interest in eating them. But decades of careful research conducted in several fields indicate that individuals do indeed vary greatly in terms of their competence for performing various tasks, and that such competence, in turn, closely reflects their skills, knowledge, motives, interests, and self-regulation (e.g., their capacity to defer rewards and stay intently focused on specific goals; e.g.,Vohs and Baumeister, 2010). So do entrepreneurs matter? The answer provided here is "absolutely"—but this perspective in no way implies that the importance of external factors will be ignored or viewed as trivial. On the contrary, the overall goal is that of presenting a balanced approach that accurately conveys the complexity inherent in the entrepreneurial process.

Are *entrepreneurs different? And if* so, why?

Now, let us turn to a slightly different, but closely related issue. Are entrepreneurs different from others persons, and if they are, why is this so? Before turning to the scientifically-valid answer, I'll begin by briefly describing some personal experiences. I am acquainted with two emergency room physicians—doctors who chose, as their specialty, working in hospital emergency rooms. One is a woman and the other a man, and they differ greatly in age and in many other ways, but they are amazingly similar in several respects. Both crave excitement and strongly dislike boredom. As one puts it "I really like it when the patients are rolling in one right after another . . . Nights when nothing happens drive me nuts!" Also, their tolerance for what many people would describe as unbearable levels of stress is amazing.

On the other hand, I also know two accountants—again a woman and a man—and they are amazingly different from my physician friends. Both are methodical, orderly, and do *not*, insofar as I can tell, like surprises! And although they experience high levels of stress around tax time, they do not crave it and both take vacations once this busy period is over.

Why do I begin by describing these individuals and differences between the physicians on the one hand, and accountants on the other? Because doing so helps illustrate a basic principle established by decades of research: *people do not choose their careers or occupations at random.* Rather, they select ones that provide an appropriate match for their own skills, preferences, motives, characteristics, and interests. Of course, many people make errors in this respect, but for most, the fit is relatively good—and if it is not, many change careers! These findings are consistent with a well-established management theory: attraction-selection-attrition (ASA) theory (e.g., Schneider, 1987).

Briefly, ASA theory suggests that the persons working in a given career or profession tend be relatively homogeneous (i.e., similar) with respect to the knowledge, skills, abilities, and other characteristics they demonstrate. The theory suggests that this similarity results from three interrelated processes: attraction, selection, and attrition. First, only some individuals are attracted to a given career or profession—for instance, only a small proportion of medical school students are attracted to becoming an emergency room physician. Second, intense selection often occurs—only individuals who find that they are well-suited for a career or occupation they find attractive actually enter it (i.e., person-organization fit; Bretz et al., 1989). Finally, only those who attain success in their chosen field remain in it while others withdraw (attrition; Ployhart et al., 2006). As a result of these interrelated processes, the persons working in any given career or occupation tend to be relatively similar to one another in certain key respects, and the longer they remain in this field, the more similar they are likely to be. In short individuals tend to choose careers or occupations to which they are initially attracted, in which their personal skills, characteristics, and motives align with the requirements of the field or career, and in which they experience a degree of success in meeting these requirements.

When this principle is applied to entrepreneurship, an intriguing possibility follows: perhaps entrepreneurs are individuals who are selected, both by their own preferences and the requirements of the entrepreneurial role, to be persons who are relatively high in certain characteristics (e.g., the capacity to tolerate high levels of uncertainty and stress; self-confidence, optimism, the desire for personal autonomy, etc.), but low in others (e.g., the desire for security and certainty, a preference for established rules and procedures). In other words, the individuals who choose to become entrepreneurs, and who remain in this career or role, constitute a select group, just as emergency room physicians, accountants, concert pianists, and world-class chess players do. Put slightly differently, many occupational or career groups differ from one another along dimensions relevant to the unique requirements of each career or profession, and the same basic principle applies to entrepreneurs (see Figure 1.3). So yes, they *are* different because the principles of attraction, selection and attrition operate for them too (Fine et al., 2012)—but not because there is anything unique or special about choosing entrepreneurship as a career.

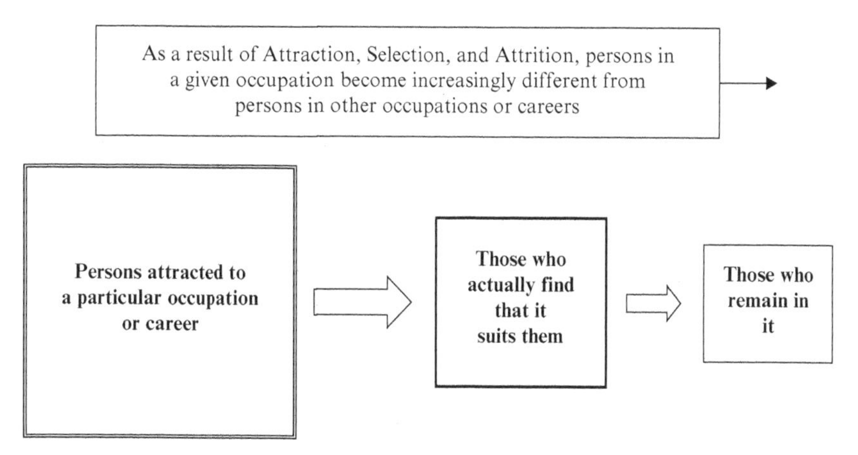

Note: Entrepreneurs, like persons in other occupational or professional groups are, as a group, highly selected. Only some are attracted to this role, only some of these persons actually become entrepreneurs, and only some of those who find that they are suited to this role remain in it.

Figure 1.3 *Are entrepeneurs different, and if so, why?*

Acquiring knowledge about the foundations of entrepreneurial success

Saying that a wide range of "tools" play a role in entrepreneurial excellence is one thing; drawing a bead on their identity is quite another. Many informal suggestions concerning the ingredients of entrepreneurial success exist, but they are simply that: untested and unverified ideas. Many may be highly insightful and even accurate, but in the absence of supporting evidence, it is risky to accept them with confidence. This raises a key question: How can we accomplish this task—or really tasks? First, how can we identify the factors that actually do play a role in entrepreneurial excellence, and second, how can we measure their relative importance? Basically, three different methods, differing greatly in approach exist: (1) asking successful entrepreneurs, (2) direct research on these issues, and (3) importing knowledge from other fields. Although they differ, they are viewed, in this book, as complementary. In other words, each provides unique kinds of information that, together, enrich our understanding of the varied components that combine to generate entrepreneurial success.

Asking entrepreneurs: can they really tell us the "secrets of success"?

At first glance, nothing seems more reasonable than asking highly successful entrepreneurs to share with us the "secrets of their success". And in fact, they are often asked precisely this: "Bill (Gates), Mark (Zuckerberg), Larry (Page)—tell us—to what do you attribute your amazing success?" When asked this question, few famous entrepreneurs, refuse to answer. And why should they? They *are* highly successful, so don't they have the right to comment on why and how they arrived at this happy state? Of course! But more importantly, can we believe them? Do they really *know* why they succeeded, when so many others failed (refer to Figure 1.1), or at least gave up along the way? Unfortunately, the answer is just as likely to be "no" as "yes", and for a simple reason: we are not nearly as good at understanding the motives behind our own behavior, or the factors that shape the outcomes we experience, as we believe. In fact, we are often wrong—dead wrong!—when we make such judgments. Memory, for instance, is fallible and easily distorted. In fact each time we recall a past event and then re-enter it in memory, it is changed—often in ways that make it more understandable, but not necessarily more accurate (Koutstahl and Cavendish, 2006). Perhaps even more disturbing, our understanding of *why* we performed certain actions in the past is uncertain and prone to error. And we are not very good at distinguishing between the factors and conditions that actually did influence either our thoughts and actions, or our outcomes, and those that did not. These are very general statements, so perhaps a concrete example will help.

In a classic experiment, one that vividly illustrates the limits of our own understanding of the factors that shape our behavior or thoughts, Nisbett and Wilson (1977) asked college students to memorize a list of word pairs. Some of these (e.g., ocean–moon) were designed to generate mental associations that would lead to certain outcomes—certain words closely related to these word pairs. For instance, participants were then asked to name a detergent. A very large proportion came up with "Tide" which is clearly related to ocean–moon. But when asked if the word pairs had influenced their answer, participants vehemently denied this; instead, they insisted that they had come up with "Tide" because they like it best, use it all the time, or because it is the most famous brand on the market. Was this true? Evidence indicates it was not, and that the word pairs had indeed affected their behavior; but the participants did not, or could not recognize this fact. In a follow-up investigation, the same researchers asked participants to choose the best product from an array of products (e.g., the best radio, pens, etc.). Results indicated that participants had a strong "right-side preference"—they chose the items on the right hand side of the display much more often than ones on the left. When asked if position had any influence on their behavior, however, they reacted with indignation: "Of course not!"

If we are not very good at even simple judgments such as these, can we be accurate in reporting the factors responsible for our success—especially since they may have existed years in the past? Please draw your own conclusion, but the weight of scientific opinion is clear: "No, we are not at all good at this task, and often make important errors." So, can we believe entrepreneurs when they tell us why and how they succeeded? Perhaps, but a degree of caution seems essential. This does not mean that we should discount such information; far from it. It is often helpful and insightful. But these statements should always be evaluated against the backdrop of evidence concerning their accuracy—evidence gathered in the ways to which we turn next. In short, we should be skeptical about the following idea, offered by one well-known writer, Tom Zelaznock (2008): "There is no simple formula for creating a successful business. Luckily, there is an easy way to improve your chances. And that's by listening to the wisdom of those who have done it already." To this we reply: good luck, because such "wisdom" is often either misleading or wrong.

Conducting systematic research

If we cannot rely on comments by entrepreneurs to solve the mystery of entrepreneurial success, how can we obtain such knowledge? Hundreds of years of scientific progress offers the following answer: "Through systematic research that collects actual data on the questions we want to address." Rather than simply asking famous and successful entrepreneurs how and why they succeeded, we can delve into these issues by studying large numbers of entrepreneurs and new ventures in order to identify the factors, actions, or strategies that are actually linked to positive outcomes (financial success or other beneficial results). Conducting such research is far from easy; we must obtain the cooperation of large numbers of entrepreneurs—individuals who are often far too busy running their companies to stop and supply the data and information we request. And interpreting the information gathered is often a very complex task. But the benefits are clear: the data obtained through research may be less subject to the biases mentioned above. Again, perhaps a concrete example might help.

It has often been suggested that entrepreneurs experience very high levels of stress; after all, they face incredibly high workloads, high levels of uncertainty, limited resources, and substantial risks. Exposure to these conditions would be expected to generate high levels of stress, and stress, a huge volume of research indicates, is often harmful to both performance and personal health. This suggests that perhaps one important ingredient in entrepreneurial excellence is the ability to resist the adverse effects of stress—and perhaps, even to thrive in its presence! How can we find out if this is true? One way is to ask successful entrepreneurs about their own ability to manage stress; but as we have noted above, the information they provide may be highly subjective and not as accurate as we would prefer.

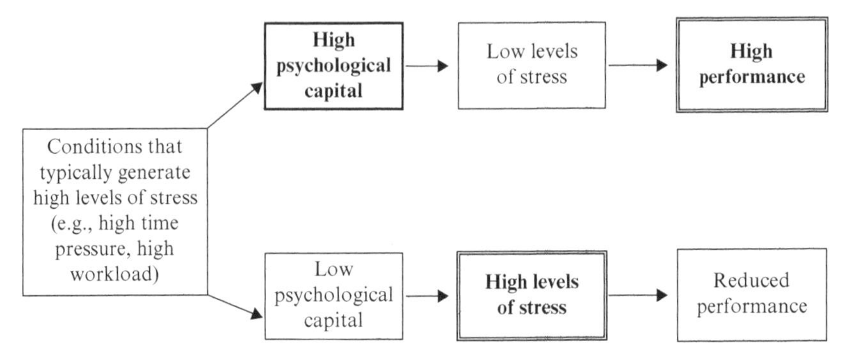

Note: Recent findings indicate that although entrepreneurs are exposed to high levels of stress, they are able to resist the adverse effects of such conditions, and that high levels of psychological capital are very helpful to them in this regard.

Source: Based on suggestions by Baron et al., 2012.

Figure 1.4 *High tolerance for stress: one essential ingredient in entrepreneurial success*

Another is to conduct research in which a large number of entrepreneurs participate by providing information on the levels of stress they experience and the extent to which they possess skills or characteristics that, we have reason to believe, might mitigate the adverse effects of stress.

One such factor is known as *psychological capital*, and involves high levels of self-efficacy (confidence in one's own ability to successfully complete various tasks), optimism, hope, and resilience—the ability to "snap back" after setbacks (Peterson et al., 2011). Previous research on psychological capital indicates that it is negatively related to stress—higher individuals are in this cluster of characteristics, the less stress they experience (Avey et al., 2011). To find out if this applies to entrepreneurs, we might ask them to complete measures of psychological capital and stress, and also obtain information on the success of their new ventures—or perhaps on the entrepreneurs' health and well-being. If psychological capital does "buffer" entrepreneurs against the adverse effects of stress, this would indicate that it is one important ingredient in entrepreneurial excellence: entrepreneurs high in this characteristic would be better able to withstand high levels of stress that might ordinarily interfere with their performance and the success of their companies (see Figure 1.4). (In fact, such research has recently been conducted, and offers support for these conclusions; Baron et al., 2012.) The higher an entrepreneur's psychological capital (as assessed by a brief but reliable questionnaire), the lower the levels of stress they report. In short, carefully conducted research is one way of gaining knowledge about the essential components of entrepreneurial excellence.

Importing knowledge from other fields

Yet one more means of obtaining information about the tools needed for entrepreneurial success involves importing such knowledge from other fields and branches of management. It seems reasonable to propose that the factors that facilitate personal success in contexts outside entrepreneurship might have similar effects in the entrepreneurial domain. In fact, such "importation" has been very useful. For instance, there is a huge body of evidence suggesting that social and political skills—a wide range of skills that help individuals to get along well with others—have a wide range of beneficial effects. For instance, people high in such skills attain greater success in their careers, have more friends and higher-quality social networks, and experience fewer interpersonal problems at work and in their personal lives, than persons lower in such skills (e.g., Ferris et al., 2007). This knowledge has been "imported" by the field of entrepreneurship, and a growing body of evidence indicates that it is directly applicable to it: entrepreneurs higher in social and political skills do obtain greater personal success than those lower in social or political skills; that is, the new ventures they found are significantly more successful, in economic terms (e.g., Baron and Tang, 2009). Many other examples of such "importation" of knowledge exist, but hopefully, this example will suffice to illustrate the value of this approach.

In sum, there are several ways of acquiring useful knowledge about the basic components of entrepreneurial excellence, and all can make unique contributions to this task. All are represented throughout the remaining portions of this book.

Tools for making the possible real: a brief overview

Before concluding this introductory chapter, we should pause and offer an overview of what will follow. In the pages that lie ahead, we will use the knowledge generated by the three methods just described to identify the key tools essential for entrepreneurial excellence—for success in making the possible real. Here is a brief overview of the major topics we will consider:

1. *Motivation*: Why do entrepreneurs undertake this demanding journey? Are they seeking fame and wealth—or something else?
2. *Creativity and Cognition*: Where do ideas for the *possible* come from? How are opportunities for entrepreneurial activities identified?
3. *Self-regulation and Self-Knowledge*: What is the role of capacities to exert self-control, stay focused, delay rewards, and know our own strengths and weaknesses?

4. *The Social side of entrepreneurship*: How can entrepreneurs build strong networks, effective teams, and develop their social or political skills?.
5. *Making effective decisions*: How can they make effective decisions, both as individuals and in groups?
6. *Personal Characteristics that Contribute to Entrepreneurial Excellence*: What personal characteristics play a role in entrepreneurial success?
7. *Dealing with Adversity and Failure*: How can entrepreneurs cope effectively with adversity and bounce back from setbacks or even failure?

That, in broad outline, is the nature of the journey we will undertake. No, we do not have a yellow brick road to guide us along the way and help us reach our ultimate goal, but we do have something even better: the thoughts, insights, and evidence provided by entrepreneurs, researchers in the field entrepreneurship, and those in several other disciplines too. Together, these sources of knowledge will help us address—and at least partially resolve—the ultimate paradox of entrepreneurship: why, among so many talented, motivated, and passionately engaged individuals, do so few actually succeed in converting the possible into the real?

Summary of key points

Entrepreneurship is not limited to the creation of new ventures. It occurs whenever and however individuals apply their creativity, ingenuity, knowledge, skills, and energy to the development of something new, useful, and better than what currently exists—something that creates some kind of value (economic, social, or other). Entrepreneurs do indeed matter; in a sense, they are the heart and soul of the entire entrepreneurial process. This view implies that their skills, knowledge, motives, interests, goals, and personal characteristics play a key role in their success. A contrasting view suggests that external factors such as when and where entrepreneurs are born, and the environments in which they grow up, are more important. However, existing evidence strongly supports the former view. Entrepreneurs, as a group, are different from persons in other occupations or careers because they represent a select sample of persons who are attracted to entrepreneurship, suited for it, and remain in it for prolonged periods of time.

Several methods exist for obtaining valid and accurate information about the "tools" essential for entrepreneurial success. These include information provided by successful entrepreneurs, systematic research, and importing knowledge about the foundations of success from other fields. All three can be informative, but emphasis in this book will be placed on the latter two because people are far less proficient at understanding the causes of their own behavior or at identifying the factors that determine the outcomes they experience. Among the topics to be examined in subsequent chapters are entrepreneurs' motives, the cognitive sources of ideas for the "possible", self-regulatory skills, social factors such as social networks and social skills, making effective decisions, personal characteristics that contribute to entrepreneurial success, and means of dealing with adversity and failure.

References

Avey, J.B., Reichard, R.J., Luthans, F., and Mhatre, K.H. (2011). Meta-analysis of the impact of positive psychological capital on employee attitudes, behaviors, and performance. *Human Resource Development Quarterly*, 22(2), 127–152.

Baron, R.A. (2012). *Entrepreneurship: An evidence-based guide*. Cheltenham, UK and Northampton, MA, USA: Edward Elgar Publishing.

Baron, R.A., and Tang, J. (2009). Entrepreneurs' social competence and new venture performance: Evidence on potential mediators and cross-industry generality. *Journal of Management*, 35, 282–306.

Baron, R.A., Franklin, R., and Hmieleski, K.M. (2012). The stress resistant entrepreneur: Why entrepreneurs flourish when others shatter. Manuscript under review.

Baumol, W. (1968). Entrepreneurship in economic theory. *American Economic Review Papers and Proceedings*, pp. 64–71.

Bretz, R.D., Ash, R.A., and Dreher, G.F. (1989). Do people make the place? An examination of the attraction-selection-attrition hypothesis. *Personnel Psychology*, 42(3), 561–581.

Ferris, G.R., Treadway, D.C., Perrewe, P.L., Brouer, R.L., Douglas, C. and Lux, S. (2007). Political skill in organizations. *Journal of Management*, 33, 290–320.

Fine, S., Meng, H., Feldman, G., and Nevo, B. (2012). Psychological predictors of successful entrepreneurship in China: An empirical study. *International Journal of Management*, 29, 279–292.

Gladwell, M. (2008). *Outliers: The story of success*. New York, NY: Little, Brown & Co.

Koutstaal, W., and Cavendish, M. (2006). Using what we know: Consequences of intentionally retrieving gist versus item-specific information. *Journal of Experimental Psychology: Learning, Memory, and Cognition*, 32, 778–791.

Nisbett, R.E., and Wilson, T.D. (1977). Telling more than we can know: Verbal reports on mental processes. *Psychological Review*, 84, 231–259.

Peterson, S.J., Luthans, F., Avolio, B.J., Walumbwa, F.O., and Zhang, Z. (2011). Psychological capital and employee performance: A latent growth modeling approach. *Personnel Psychology*, 64, 427–450.

Ployhart, R.E., Weekley, J.A., and Baughman, K. (2006). The structure and function of human capital emergence: A multilevel examination of the Attraction-Selection-Attrition model. *Academy of Management Journal*, 49, 661–677.

Schneider, B. (1987). The people make the place. *Personnel Psychology*, 40, 437–453.

Shane, S. (2012). Reflections on the 2010 AMR decade award: Delivering on the promise of entrepreneurship as a field of research. *Academy of Management Review*, 37, 10–20.

Shane, S., and Venkataraman, S. (2000). The promise of entrepreneurship as a field of research. *Academy of Management Review*, 25, 217–226.

Vohs, K.D., and Baumeister, R.D. (2010) (eds). *Handbook of Self-regulation*, 2nd edn. New York, NY: Guilford Press.

Zelaznock, T. (2008, 29 February). 7 entrepreneurs whose perseverance will inspire you. Growthing Blog.

2 Motivation: why entrepreneurs make the journey

Chapter outline

Human motivation: energizing, guiding, and maintaining goal-directed actions
The nature of human motivation: contrasting views
Expectancy theory: the "pull" of future events
Goal-setting theory: the "pull" of specific objectives
Prelude to action: entrepreneurial intentions
What do entrepreneurs really seek? Wealth? Fame? Doing social-good? Answer: all of the above
Why motivation is an important tool for entrepreneurs

* * *

> Happiness is not in the mere possession of money; it lies in the joy of achievement, in the thrill of creative effort.
> (Franklin D. Roosevelt)

> If you work just for money, you'll never make it, but if you love what you're doing and you always put the customer first, success will be yours.
> (Ray Kroc)

What comes to mind when you hear the word "entrepreneur?" For most people, this term conjures up images of successful, famous individuals (Steve Jobs, Bill Gates, Ron Schaich, founder of Panera Bread). And closely associated with these images are thoughts of the vast wealth they have acquired—wealth that allows them to live any kind of life they choose. But think for a moment: what would *you* do if you had $1 billion, or even $100 million? Continue to work? Begin a perpetual vacation? Move to some beautiful spot? In fact, when asked to imagine what they would do if they suddenly possessed huge wealth, most people say something like this: "I don't know for sure, but I *do* know one thing: I'd never do anything I didn't *want* to do again." And then many continue by outlining plans that are surprisingly subdued and far from dramatic: "I guess I'd buy some things for myself and for my family—a new house, a new car, or maybe several . . . but I'd probably want to keep living pretty much the way I do, and probably even keep on working." In fact, that is exactly what many winners of giant lottery prizes do: after a short period of treating themselves to the pleasures of fulfilling long-held wishes, they go back to work—often, to the same (or highly similar) jobs. While this might seem surprising, it meshes closely with research indicating that when asked how happy they are with their lives, most people (85 percent or more!) report that they are either very or at least moderately happy (e.g., Diener et al., 2010). Moreover, this seems to be true in relatively poor countries as well as rich ones. In fact, people in poor nations tend to report being just as happy—and in some cases, happier—than in rich ones.

More directly related to this discussion, the same general pattern seems to prevail for successful entrepreneurs. While some do "retire" from active life once they attain personal wealth (I know one who purchased a yacht and now, ten years later, has been sailing the world ever since!). Many others, however, prefer to continue working—and persist in seeking new challenges. Years ago, for instance, I knew an entrepreneur who had recently sold his company for almost $1 billion. When I asked him, "What will you do now?" He did not hesitate a second in answering: "Start another company—in fact, I already have." So, consistent with the quotations at the start of this chapter, money does not seem, in and of itself, to be a source of happiness, and people—and especially entrepreneurs—who attain it, often do not take the time to enjoy the opportunities it provides.

These findings and observations raise an intriguing question: What, precisely, do entrepreneurs seek in their efforts to launch and develop new ventures? It is all-to-easy to assume that their primary goal is wealth, but is that really true? And if they are not seeking wealth, what other benefits do they hope to enjoy from converting their dreams and visions—the *possible*—into the real? That is the central issue on which we will focus in this chapter. In order to provide you with insights into this complex issue—and, perhaps, into your own motives and goals for becoming an entrepreneur—we will proceed as follows.

First, we will examine the basic nature of human motivation—what it is and the central role it plays in our lives. Second, we will examine the link between motivation and overt actions—a route that involves concrete *intentions*. Third, we will focus on intentions—the cognitive mechanism through which motives are translated into overt actions. Finally, and to paraphrase words used in Chapter 1 ("beyond new ventures") we will look beyond personal wealth, at the other major motives that underlie entrepreneurs' actions. If you believe that most entrepreneurs are strongly motivated to and focus strongly on the attainment of personal wealth, get ready for some surprises, because existing evidence indicates that this is probably not the case.

Human motivation: energizing, guiding, and maintaining goal-directed actions

People often engage in actions for which no immediate or obvious (some would say, rational) cause is obvious. For instance, rock climbers edge their way up sheer rock walls from which a fall would result in serious injury or even death (see Figure 2.1). Persons seeking political change take to the streets where, with few weapons, they find themselves confronting tanks and a huge arsenal of modern weapons controlled by the current regime (as, tragically, in 2011 and 2012, in Syria). And still others collect stamps, coins, or other items and devote great effort and a large portion of their wealth, to obtaining the rare items their collections lack. Finally, consider people who choose to leave secure and well-paying jobs, positions that provide health insurance and many other benefits, for a field in which the odds are stacked highly against them—to become entrepreneurs!

In all these instances, we find ourselves asking a basic question: *Why*? What factors or conditions lead the people in question to engage in activities that, to many others, seem puzzling in the extreme. The answer offered

Note: Why do people engage in activities such as the one shown here? Motivation, which refers to unobservable, internal processes that energize, guide, and maintain behavior, provides an answer. Motivation helps explain behavior when no obvious external causes for its occurrence are available.

Source: Fotolia 24354841.

Figure 2.1 *Motivation: the internal processes that energize, guide, and maintain human behavior*

by psychologists and others who delve into the mysteries of human behavior is that these actions stem from a basic process that plays a key role in our thoughts and actions: *motivation*. This term refers to internal processes (cognitive emotional, biological) that cannot be directly observed, but that, nevertheless, exert powerful effects on what individuals think, plan, and do. Since these internal processes cannot be directly seen (although they can often be measured), their presence must be largely *inferred*—for instance, we observe others' actions and from these, infer that they must be pursuing various motives. Whatever its precise nature or origins however, motivation has three crucial effects: (1) it provides the energy or impetus for overt actions, (2) it guides them, and (3) it causes them to persist until the motive is met or fulfilled. For instance, rock climbers seek out higher and higher cliffs, expend lots of energy climbing to the top, and continue in this activity for days, months, or years. The same is true of sky-jumpers, stamp collectors and many other groups that engage in actions many people find difficult to understand (see Figure 2.1). The same basic principles apply to entrepreneurs. Generally, they must exert large amounts of effort and energy to carry out the steps needed for starting a new venture—or for acting entrepreneurially in other ways (e.g., within large companies, in their own professional practices, etc.). In addition, these actions are generally directed toward specific goals. For instance, an entrepreneur seeking funding will not (we hope!) visit a casino to obtain the financial resources he or she needs by gambling. Rather, the entrepreneur will approach banks, venture capitalists, or business angels to obtain the needed funds. And third, these efforts will persist; in fact, they may well continue over several months or even years. In short, motivation often plays a key role in human behavior and helps us understand—and explain—actions for which there are no obvious external causes.

Suggesting that entrepreneurial activities involve motivation, however, does not answer another fundamental question: What is the nature of this motivation? For rock climbers, it might be described as the desire to experience excitement or danger. For political protestors, it may involve the motivation for greater personal freedom. But what is the motivation for entrepreneurs—what internal processes lead them to risk their careers, reputation, personal wealth, and even the well-being of their families, in activities that offer little guarantee of positive outcomes (recall the failure rate of new ventures presented in Chapter 1)?

This is the basic issue on which we will focus in this chapter. In order to address this question carefully, however, we must complete preliminary steps that lead us toward a useful answer. First, we will briefly consider contrasting views about the nature of motivation—the basic processes that underlie it. Then, we will examine intentions, cognitions that link motivation to overt actions. Finally, we will focus on the motives that lead entrepreneurs to their vigorous and often persistent efforts to create something new and better.

The nature of human motivation: contrasting views

As noted earlier, motivation cannot, in general, be observed directly, but its effects on human behavior *are* observable. In fact, motivation plays a key role in any activity requiring persistent, concentrated effort. This raises an intriguing question: What is its basic nature? Even if we cannot observe motivation itself, perhaps we can understand the foundations from which it derives. The results of decades of research on this issue indicate that, as you might guess, motivation springs from several different sources. At the most basic level, biological factors play a role. For instance, hunger and thirst reflect basic bodily requirements, and everyone knows from personal experience that they can powerfully energize, direct, and maintain behavior. When hungry, individuals focus intently on meeting this need, and will go to great lengths and expend great effort to meet this motive. More important for our present discussion, however, are two other views of the foundations of motivation that relate more directly to cognition—the "thinking" side of life.

Expectancy theory: the "pull" of future events

In many instances, our current behavior is determined not by biological needs, but by our thoughts. Basically, our behavior at the present time is influenced by our thoughts about future events and outcomes. This basic reasoning underlies *expectancy theory*, a highly influential view of human motivation.

According to this perspective, three factors form the basis for motivation—and especially for motivation to work hard and expend effort on various tasks. The first is *expectancy*, the belief that effort will result in improved performance. The second is *instrumentality*, the belief that improved performance on key tasks will lead to desired rewards, whatever these are. The third, *valence*, refers to the value individuals place on these rewards. The theory further proposes that motivation is a multiplicative function of these three factors: expectancy x instrumentality x valence. The upshot is that people will work hardest and most persistently on various tasks when they believe that expending effort will improve performance, enhanced performance will lead to desired outcomes, and these outcomes themselves are ones they truly want.

The broad outlines of expectancy theory have been verified by decades of research, so overall, it helps explain why individuals—including entrepreneurs—sometimes invest countless hours of hard, and often tedious, work on specific tasks. Essentially expectancy theory suggests that they are willing to do so because they believe that these actions will help them obtain the rewards or goals they strongly desire. A key implication of this theory is that entrepreneurs, who often invest almost limitless time and energy in their new ventures, are doing so because they seek certain outcomes. We will examine these outcomes—which are very varied in scope—in detail in a later section, but here, want to emphasize the "future-driven" approach of this perspective on human motivation.

Goal-setting theory: the "pull" of specific objectives

Another, and in some ways, closely related theory of motivation is known as *goal-setting theory*. This theory, too, emphasizes cognitive factors but in this framework, the focus is directly on concrete goals (e.g., Locke and Latham, 2002). According to this theory, individuals identify various goals (ones they choose themselves or ones assigned to them by, for instance, a supervisor, teacher, or parent). Then, they direct their effort and actions toward reaching these goals. A vast amount of research evidence indicates that setting certain kinds of goals is, perhaps, the best single means for generating high levels of effort and persistence, and hence, for improving performance on a wide range of tasks. However, additional evidence indicates that goal setting works best under certain conditions. It is most effective in boosting performance when the goals set are highly *specific*—people know just what they are trying to accomplish, the goals are *challenging*—meeting them requires considerable effort, but they are perceived as *attainable*—people believe they can actually reach them. Finally, goal setting is most successful when individuals receive feedback on their progress toward meeting the goals and when they are truly and deeply committed to reaching them. This last point is quite important; if goals are set by someone else and the people who are expected to meet these goals are not committed to doing so, then goal setting can be totally ineffective, and may even backfire, reducing rather than enhancing motivation. When the conditions described above are met, though, goal setting is a highly effective way of increasing motivation and performance (e.g., Locke and Latham, 2002).

Think for a moment about the implications of these facts for entrepreneurs. As we will note in more detail below, many choose this career or activity because they strongly desire autonomy—they want to be able to set their own goals, rather than have someone set them. That, in turn, suggests that entrepreneurs may differ from many other groups in terms of certain motives—again, a point that is central to this chapter, and that we will examine in more detail shortly. (An overview of expectancy theory and goal-setting theory is presented in Figure 2.2.)

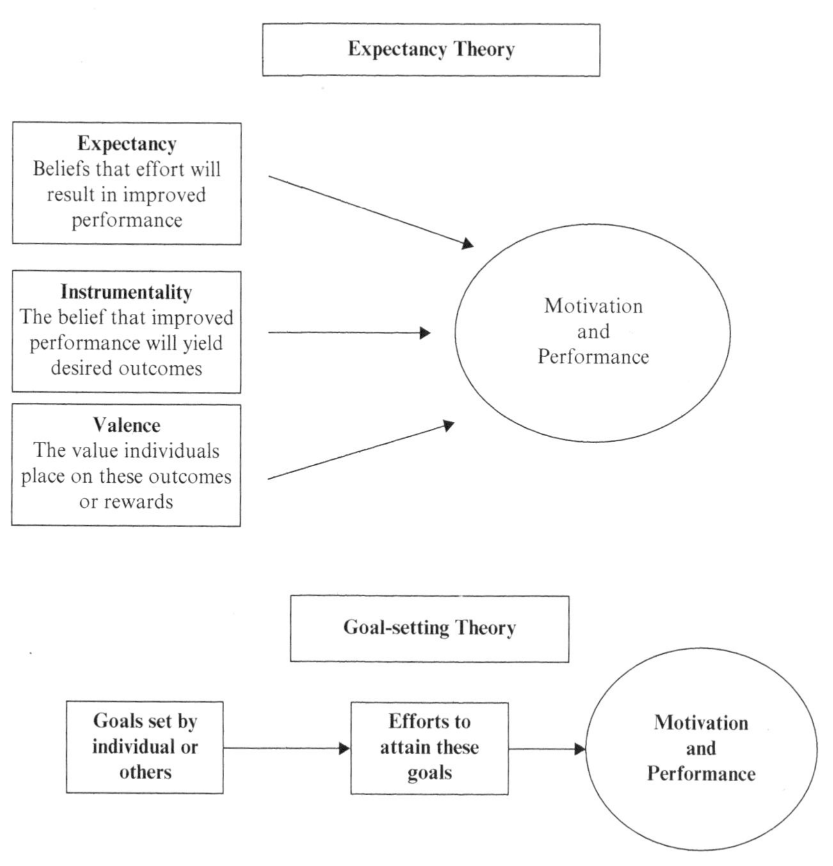

Note: Both of these theories emphasize the importance of cognitive factors—an individual's thoughts about future events or outcomes. Expectancy theory suggests that motivation (and performance) is a function of expectancy, instrumentality, and valence. Goal-setting theory, in contrast, focuses directly on goals and efforts to obtain them.

Figure 2.2 *Expectancy theory and goal-setting theory: views of human motivation*

To conclude: although we cannot see it, touch it, or feel it, motivation is a very important force in human behavior. In fact, it is fair to say that without it, nothing much in the way of persistent, directed, behavior occurs. It is only when individuals perceive various actions as effective means for attaining desired rewards or goals that they become engaged, so motivation is indeed a key ingredient in performance in many different contexts, and is certainly a crucial component in entrepreneurial success. One final point: motivation is not, in any sense, a fixed or immutable process. On the contrary, it can be increased, reduced, strengthened, or depleted by both external conditions (e.g., a highly unreasonable boss) and internal factors (e.g., beliefs by individuals that they can successfully accomplish tasks they set out to accomplish—self-efficacy; e.g., Bandura, 1997, 2012).

Prelude to action: entrepreneurial intentions

Do motives lead directly to action? In some cases, they do. A thirsty individual may focus strongly on obtaining a drink, and engage in actions needed to satisfy this powerful biological need. But for more complex motives, such as the ones that underlie entrepreneurial action, motives are often not translated directly into action. Rather, they lead to intervening cognitive processes that play a key role in whether, and how, the motives in question are expressed. Recognizing this fact, much research on the question of why individuals choose to become entrepreneurs has focused not on their underlying motives (wealth, self-fulfillment, doing social good), but rather on their *intentions*—the concrete plan to become an entrepreneur (e.g., Zhao et al., 2010).

What, specifically, are intentions, and how do they exert their effects on behavior? Answers are provided by two closely related theories: *the theory of reasoned action* and *the theory of planned behavior* (Ajzen, 1987, 2001; Ajzen and Fishbein, 2005). Both theories define behavioral intentions as an individual's readiness to perform a given behavior—a readiness that is often an immediate antecedent of that behavior. In fact, extensive research indicates that intentions are, perhaps, the best single predictor of specific actions (e.g., Baron and Branscombe, 2012). For instance, consider voting behavior. An individual's concrete intention to vote for a particular candidate is usually a much better predictor of her or his actual vote than, for instance, that person's general attitude toward voting as a social responsibility or her or his attitudes toward a particular political party.

How, specifically, do intentions influence behavior? According to both theories mentioned above the decision to engage in a particular behavior is the result of a rational process. Various options are considered, the consequences or outcomes of each are evaluated, and a decision is reached to act or not act. That decision is then reflected in *behavioral intentions*, which, as already noted, are often very accurate predictors of whether the behaviors will actually be performed (Ajzen, 1987; Albarracin et al., 2011).

Interestingly, the link between intentions and overt behavior is especially strong when people have formed a plan for how and when they will translate their intentions into behavior (Frye and Lord, 2009; Webb and Sheeran, 2007). In this sense, preparing a business plan, or even developing a clear business model showing how an opportunity might be developed into a profitable business, can be highly useful for entrepreneurs. By formulating plans for implementing intentions, entrepreneurs develop specific procedures for translating their motives (e.g., achieving financial independence) into concrete actions. Intentions, themselves, are strongly shaped by two factors: *attitudes toward the behavior*—people's positive or negative evaluations of the behaviors in question (e.g., whether they are expected to yield positive or negative consequences), and *subjective norms*—perceptions of whether others will approve or disapprove of this behavior. A third factor: *perceived behavioral control*—people's appraisals of their ability to perform the behavior—was subsequently added to the theory (Ajzen, 1991).

The implications of this widely-accepted perspective for the field of entrepreneurship are clear: as we saw in Chapter 1, most persons hold very positive attitudes toward entrepreneurship, and in many societies, engaging in entrepreneurship is perceived as a worthwhile and valuable activity. Further, research findings indicate that entrepreneurs are far above average in self-efficacy—the belief that they can successfully perform the activities they undertake (e.g., Bandura, 2012). Thus, they are high in perceived behavioral control—they believe that they can, in fact, successfully perform the actions involved in starting a new venture of other forms of entrepreneurship. Overall, then, the theory of planned behavior suggests that the intention to become an entrepreneur will be closely related to actually becoming one.

As noted above, researchers in the field of entrepreneurship have taken careful note of this fact, with the result that most research on entrepreneurs' motivation—the question of *why* they become entrepreneurs in the first place—has focused on their intentions to adopt this role or career. More specifically, this research has often sought to determine if certain personal characteristics influence such intentions, and therefore, the likelihood that individuals will become entrepreneurs. Strong evidence for such relationships has recently been reported by Zhao et al. (2010), who analysed the data from many studies that had previously investigated the relationships between personal characteristics (especially, certain key aspects of personality) and *entrepreneurial intentions*, intentions to undertake entrepreneurial actions.

Zhao et al. (2010) found that several basic aspects of personality (e.g., Costa and McCrae, 1992) are indeed significantly linked to entrepreneurial intentions. In other words, the greater the degree to which individuals demonstrate certain personal characteristics, the more likely they are to form the intention of becoming an entrepreneur. These characteristics include: (1) *conscientiousness*—the tendency to be high in achievement and work motivation, organization and planning, self-control, responsibility, (2) *openness to experience*—the tendency to be high in curiosity, imagination, creativity, seeking our new ideas; and (3) *emotional stability*—tendencies to be calm, stable, even-tempered, and hardy (i.e., resilient). All of these characteristics were positively related to entrepreneurial intentions. In contrast, another basic dimension of personality, *extraversion*—being outgoing, warm, friendly, energetic, and assertive—was only weakly related to such intentions. Finally a fifth dimension—*agreeableness*—which involves being trusting, altruistic, cooperative and modest—was not linked to such intentions. The tendency to take risks was also positively related to entrepreneurial intentions, so as has often been proposed; persons who become entrepreneurs are more willing than others to enter situations involving high degrees of uncertainty concerning success or failure.

Although the aspects of personality, and risk-taking propensity, play an important role in entrepreneurial intentions, other individual characteristics, too, are relevant. Among these *self-efficacy*—individuals' belief that they can successfully accomplish tasks they choose to undertake (Bandura, 1997)—is perhaps the most important (e.g., Zhao et al., 2005). Research findings indicate a positive relationship between self-efficacy and both the tendency to actually start new ventures (Markman et al., 2003, 2005; Zhao et al., 2005; average correlation = 0.25) and the financial success of these businesses (Chandler and Jansen, 1991). Another is achievement motivation (or need for achievement). This refers to the desire to meet or exceed high standards, and attain excellence. Not surprisingly, entrepreneurs tend to be high on this dimension—higher, by far, than most other persons (e.g., Collins et al., 2004).

In sum, it appears that some individuals are indeed more likely than others to form the intention of becoming an entrepreneur. But what basic motives underlie these intentions? That topic is at the very heart of this discussion, and it is the one to which we turn next.

What do entrepreneurs really seek? Wealth? Fame? Doing social-good? Answer: all of the above

Why do individuals choose to become entrepreneurs? Why do they take the concrete steps needed to convert their ideas or dreams to reality, either by starting new ventures or through other means? The fact that acting entrepreneurially involves high levels of uncertainty and no guarantees of success, suggest that this choice must involve strong underlying motives—internal processes we cannot directly observe, but, what, precisely, are these? It has often been assumed that the answer to this question is straightforward: entrepreneurs primarily seek wealth or fame, and view starting a new venture (or other entrepreneurial actions) as useful for attaining these goals. All too often, this is the implicit message in courses in entrepreneurship. Examples and cases emphasize highly successful new ventures and the benefits they provide for their founders. Further, business plan competitions award substantial cash prizes to the winners, and—perhaps even more important—provide the entrepreneurs with introductions and access to venture capitalists, business angels, and other potential sources of funding. The overall result is that the term "entrepreneur" is now strongly associated, both in the public mind and in the field of entrepreneurship itself, with strong motivation for financial gain. Yes, the possibility that entrepreneurs seek other goals and proceed from different motives, is mentioned occasionally—especially in the context of social entrepreneurship—which is viewed as deriving, primarily, from motivation to create not wealth, but social good, beneficial outcomes for a large number of persons without, necessarily, generating large financial rewards for the entrepreneurs. By and large, however, the decision to become an entrepreneur is often perceived as motivated by strong desires for success, fame, and personal wealth.

Is this view accurate? Although it may apply to many current or future entrepreneurs, it tends to downplay or ignore other motives that may be just as strong—or even stronger—than financial ones. For example, some individuals become entrepreneurs largely out of necessity rather than a desire for fame and wealth: they are downsized by their employers or lose their current jobs for other reasons, but are too young to retire—or simply are not ready to do so. If they are over the age of 50, they may be viewed as relatively undesirable in the job market, so such persons often turn to entrepreneurship as one possible option.

Despite growing recognition of such facts, however, wealth creation is still widely viewed as the primary motivation for entrepreneurship. In an influential paper, however, several well-known scholars (Rindova et al., 2009) propose, very clearly, that this view greatly misses the mark. While they recognize the importance of economic motives in entrepreneurship, they suggest that in fact, many individuals engage in such activities because they are seeking what they term *emancipation*—release from social economic, cultural or institutional constraints that restrict their personal freedom or autonomy. Consistent with this view, they propose use of the term *entrepreneuring*, which they define as efforts to bring about new economic, social, institutional, and cultural environments. In other words, entrepreneurs often seek emancipation through change, and act to produce such shifts or transformations.

As a prime example, Rindova et al. (2009) point to Google, noting that its founders wanted to break free from constraints on use of the Internet that existed in the mid-1990s. Their goal was to shatter technological and cultural barriers, so that vast amounts of information would be available to virtually anyone. It is important to note that at the same time they also wanted to build a profitable company; as Rindova et al. (2009, p. 483) note, "there is no opposition between emancipatory projects to create change and a hard-nosed business strategy". In fact, entrepreneurship—like all complex behavior—stems from multiple motives, often operating simultaneously. What is crucial, however, is to recognize that in many instances, wealth generation is not the only, or even most important one.

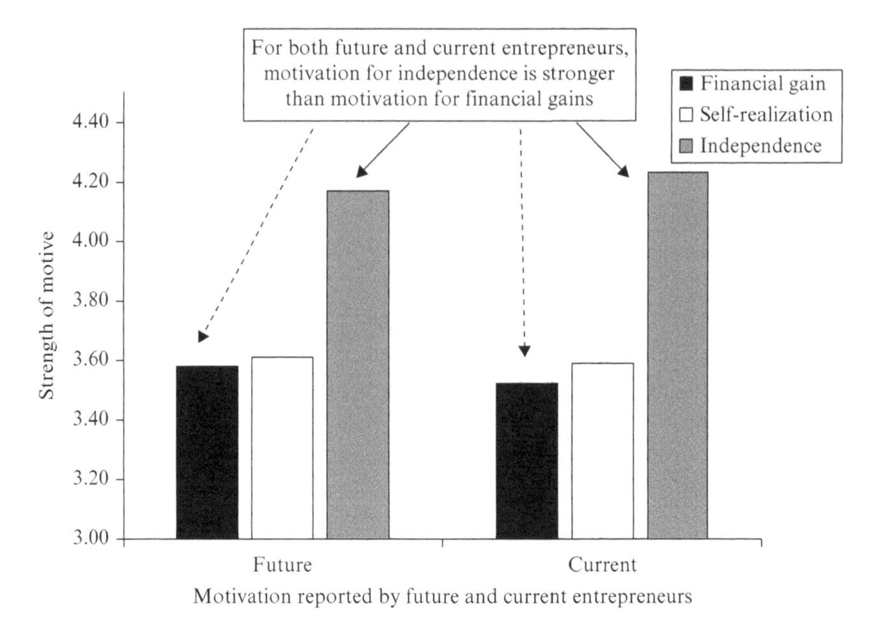

Note: Growing evidence indicates that entrepreneurs have many different motives and that among these, the desire for independence or self-fulfillment is often stronger than the desire for financial success.

Source: Based on data from Cassar, 2007.

Figure 2.3 *Evidence that entrepreneurs seek more than simply financial success*

Another way to express this view is to note that many entrepreneurs start new ventures or perform other entrepreneurial actions because they are seeking self-realization. They want to gain to grow and learn as a person, to exert leadership, motivate others, or fulfill personal visions and dreams. A different, but somewhat related motive, involves a desire for independence—to gain personal autonomy and, as Rindova et al. (2009) suggest—break free from constraints. How important are such motives? Evidence provided by a study conducted by Cassar (2007) indicates that they are very important. In this investigation data from a large national survey of entrepreneurs and prospective (nascent) entrepreneurs (the PSED, which includes more than 64,000 entries!), were used, as a basis for evaluating the relative strength of several motives that could, potentially, underlie entrepreneurship. Several different motives were also assessed, and as shown in Figure 2.3, results were clear: the desire for independence was stronger than the desire for financial success among both prospective and actual entrepreneurs. Still another motive, the desire for recognition, which is closely related to motivation for achievement, was also fairly high in the ratings.

Similarly, it has recently been proposed by the present author (Baron, 2010) that individuals often become entrepreneurs because they are seeking the kind of working conditions identified by research on job design as ones that generate high levels of motivation, satisfaction, and performance (e.g., Baron, 2010; Fried and Jullerat, 2011). In other words, they become entrepreneurs because they are seeking *meaning* in their work, and this is provided by high levels of autonomy, task significance (tasks being performed seem important), task identity (entire tasks can be completed, not simply portions of them), and skill variety (many different skills can be exercised and put to good use). Clearly, the entrepreneurial role provides these conditions, and it may be that individuals who perceive that they can attain such conditions by starting their own businesses or acting entrepreneurially in other ways, are especially attracted to this role. In essence, they become entrepreneurs in a quest for "the perfect job or career"—one that will be closely aligned with their own strong desires for meaningfulness in what they do, and in this way, will contribute to their personal happiness. For such persons, the decision to become an entrepreneur is certainly *not* the result of strong desires for wealth and fame. Nor are these the key factors for individuals often known as "life style entrepreneurs". These are persons who find the idea of being an entrepreneur, with the aura of being a "free and independent spirit" to be highly appealing. For instance, I formerly had the pleasure of team-teaching a course on entrepreneurship with a Nobel-prize winning physicist. He had started a company to develop his scientific ideas into useable products (one that could diagnose cancer on the basis of the electrical properties of cells). But he took virtually no steps to obtain adequate financing or build sales. Rather, he seemed to enjoy working in the University's Incubator Park, where as Nobel-scientist, he was always welcome and always treated with great respect (deservedly so). He was certainly not motivated by a desire for wealth or fame—he already had as much fame as most people could ever want, and was reasonably well-off financially, too. But he did seem to immensely enjoy the role of "scientist-turned-entrepreneur" and it appeared to contribute to his sense of personal fulfillment and well-being.

In addition, some persons become entrepreneurs because they want to "do good"—to enhance human well-being, or at least, the well-being of people in their own communities, regions, or countries. Such persons are often described as being social entrepreneurs, and although they realize that they must make a profit to stay in business and to continue actions that will be beneficial to many other persons, they are not focused on this goal. For example, consider a company called Sun Catlytix, founded by Professor Daniel Nocera, of MIT. Professor Nocera has developed an invention he describes as an "artificial leaf"—a device consisting of a sheet of silicon with different catalytic materials (cobalt, nickel) bonded to its two sides (see Figure 2.4). When placed in ordinary water and exposed to sunlight, the "artificial leaf" splits the water molecules into hydrogen and

Note: The "artificial leaf", a device that can harness sunlight to split water into hydrogen and oxygen without needing any external connections, is seen with some real leaves, which also convert the energy of sunlight directly into storable chemical form. The hydrogen it produces can then be used as a fuel by huge numbers of persons in underdeveloped countries who are not on existing power grids. The device was developed by Professor Daniel Nocera of MIT.

Source: Courtesy of MIT News Office. Photograph by Dominick Reuter.

Figure 2.4 *Providing the world with a renewable, non-polluting source of energy*

oxygen, just as natural leaves do during photosynthesis (Owen, 2012). The hydrogen generated can be used as fuel, and the possibilities for this device are nothing short of amazing: Professor Nocera estimates that just three gallons of water could provide enough energy (hydrogen) to power a large American household for an entire day. Certainly, providing products that can accomplish this feat can be highly profitable, and Sun Catlytix, like all companies, must generate profits to stay in business. But as Professor Nocera describes it, the key goal is not necessarily financial gain, but providing billions of people in undeveloped countries with an inexpensive, reliable, and non-polluting source of power—the equivalent of about 100 watts. In short, although large profits might well result, Sun Catlytix is focused on helping to solve what Professor Nocera sees as the fundamental problem facing humanity—continued population growth. In his own words: "If I give poor people energy, they become empowered, and every study that's ever been done has shown that with financial gain and education, population growth drops like a rock."

In sum, individuals become entrepreneurs for many reasons, and gaining personal wealth is only one of these. Desires for self-fulfillment, personal growth, and helping others are often as important, or more important, than motivation for financial gain. It is important to quickly add that there is no intention here of suggesting that one of these motives is superior or more appropriate than the others. In fact, decades of research on human motivation suggest that complex actions rarely reflect a single motive or goal. Rather, several often operate simultaneously, and it seems very likely that this is true for the decision to become an entrepreneur, or simply to act entrepreneurially. Further, it is also important to note that new ventures *must* have cash flow and be profitable if they are to survive, even if their primary goals are not financial in nature. There is one potential "downside" to strong motivation for financial gain, however. Recent research (Baron et al., 2012) indicates that the stronger this motive, the more likely entrepreneurs are to "cut corners" or even engage in unethical actions. In contrast, there is no similar relationship between the desire for self-fulfillment and such actions. This suggests one reason why, perhaps, the field of entrepreneurship might at least consider reducing its emphasis on the financial benefits of becoming an entrepreneur.

Before concluding this discussion, it is important to mention one additional point: individual motives, characteristics and preferences are certainly important in the decision to become an entrepreneur, but they are definitely not the entire story. As we noted in Chapter 1, external (i.e., societal, social) factors, too, often play a role. For instance, individuals are more likely to vigorously pursue entrepreneurial actions when financial markets are favorable so that access to essential financial resources is readily available, when government policies favor the founding and operation of new ventures (e.g., through favorable tax policies; Tung et al., 2006), and when markets for specific new products or services are growing rapidly (Shane, 2008). These factors, too, are important because individuals do not choose to pursue their dreams through entrepreneurial activity in isolation from social and economic conditions. Trying to understand entrepreneurship without considering these factors is like trying to prepare a complex recipe without a full list of the key ingredients.

Why motivation is an important tool for entrepreneurs

The central theme of this book is identifying the tools entrepreneurs need to succeed—the personal equipment that can help them effectively convert their dreams of the *possible* into something real. Is motivation one of these tools? In one sense, it can be viewed as a basic requirement: without it, there is no "engine" and nothing else follows. In another sense, though, it is indeed a tool that can be applied to many aspects of entrepreneurial activity. Consider the following saying: "It is hard to get somewhere unless you know where you want to go." Applied to motivation, this suggests that unless entrepreneurs are clear on *why* they are starting a new venture or engaging in other entrepreneurial actions, the engine will be present but another essential component—the guidance system—will be lacking, or at least impaired. In other words, entrepreneurs will be less able to shape their actions, strategies, and plans so as to maximize the odds of reaching the goals they seek. Are they seeking personal wealth? Then one set of actions and strategies may be most effective. Are they seeking, instead, to maximize their autonomy? Then another—for instance, choosing to provide their own financing rather than obtain it from venture capitalists or others—may be preferable. Are they seeking to "do social good"—to contribute to the well-being of large numbers of other persons? Then still other strategies may be most appropriate.

In short, motivation is important in many respects, but to use it in the most effective ways, entrepreneurs must gain insight into what they, personally, seek to accomplish—what specific motives are most important to them personally. To the extent they do, not only will they depart on an exciting, challenging, but potentially rewarding journey—they will, like Dorothy, know what they want to achieve. And that, even in the absence of a yellow brick road, will help them plan a route to their final destination.

Some closing thoughts

To conclude: although entrepreneurs clearly differ greatly in terms of motives, interests, values, skills, personal characteristics and many other factors, they do share one basic attribute: they are all active *doers*—persons who do not merely dream or envision, but instead, take action to convert these ideas into reality. For this reason, motivation is indeed central to entrepreneurial success. As Locke and Baum (2007, p.93) put it: "A person may have sufficient technical skill and money to start a business, but without motivation [to do so] *nothing happens*." In other words, entrepreneurs are indeed the "active principle" in entrepreneurship, and it is their efforts—often intense and prolonged—to reach their goals and fulfill their dreams that makes the entire process go.

Summary of key points

Why do entrepreneurs engage in intense, prolonged efforts to create something new, useful, and better? Most people seem to assume that they do so in order to gain wealth and personal fame, but current knowledge of human motivation—the internal processes that energize, guide, and maintain human behavior—suggest that this view may be too unitary in nature. In fact, entrepreneurs seek a multitude of motives, including personal wealth, greater independence, and even helping others.

Motivation is a useful construct in situations where individuals engage in actions for which no obvious external causes can be identified. Rock climbers, sky-divers, avid collectors (stamps, coins, whatever) engage in such actions. For such persons, motivation to experience excitement or stimulation may be key. Many explanations of the nature of human motivation exist, but among these, two that are very influential—expectancy theory and goal-setting theory—emphasize the idea that cognitive factors play a key role. In essence, individuals engage in many actions because they believe that doing so will help them attain outcomes they desire. Motivation does not directly influence behavior; rather, it often underlies behavioral intentions—a readiness to perform a given action. Intentions are, perhaps, the best single predictor of overt action. Intentions to become an entrepreneur are related to several different and very basic personal characteristics (e.g., conscientiousness, openness to experience, emotional stability), as well as self-efficacy and achievement motivation.

Research on the question "What do entrepreneurs really seek"—in other words, what motivates them?—indicates that the desire for personal wealth is often less important than desires for meaningful work, independence, or self-realization. In addition, they may seek "emancipation" from current technological, social, or cultural constraints. And some, social entrepreneurs, seek, primarily, to do "social good"—to help large numbers of persons in their communities, societies, or even the entire world. In short, assuming that entrepreneurs are focused on attaining fame and wealth is misleading, and does not accurately represent the wealth of factors that induce them to undertake efforts to make the possible real.

References

Ajzen, I. (1987). Attitudes, traits, and actions. Dispositional prediction of behavior in personality and social psychology. In L. Berkowitz (ed.), *Advances in Experimental Social Psychology*, Vol. 2, pp. 267–299. New York: Academic Press.

Ajzen, I. (1991). The theory of planned behavior. *Organizational Behavior and Human Decision Processes*, 50, 179–211.

Ajzen, I. (2001). Nature and operation of attitudes. *Annual Review of Psychology*, 52, 27–58.

Ajzen, I., and Fishbein, M. (2005). The influence of attitudes on behavior. In D. Albarracin, B.T. Johnson, and M.P. Zanna (eds), *The Handbook of Attitudes*, pp. 173–221. Mahwah, NJ: Lawrence Erlbaum.

Albarracin, D., Johnson, B.T., Fishbein, M., and Muellerleile, P.A. (2011). Theories of reasoned action and planned behavior as models of condom use: A meta-analysis. *Psychological Bulletin*, 127, 142–161.

Bandura, A. (1997). *Self-efficacy: The exercise of control*. New York: W.H. Freeman.

Bandura, A. (2012). On the functional properties of perceived self-efficacy revisited. *Journal of Management*, 38, 9–44.

Baron, R.A. (2010). Job design and entrepreneurship: Why closer connections = mutual gains. *Journal of Organizational Behavior*, 30, 1–10.

Baron, R.A., and Branscombe, N.R. (2012). *Social Psychology*, 13th edn. Upper Saddle River, NJ: Pearson.

Baron, R.A., Zhao, H., and Qing, M. (2012). Personal motives and unethical behavior among entrepreneurs: Cognitive mechanisms that lubricate the "slippery slope". Manuscript under review.

Cassar, G. (2007). Money, money, money? A longitudinal investigation of entrepreneur career reasons, growth preferences, and achieved growth. *Entrepreneurship & Regional Development*, 19, 89–107.

Chandler, G.N., and Jansen, E. (1991). The founder's self-assessed competence and venture performance. *Journal of Business Venturing*, 7, 223–236.

Collins, C.J., Hanges, P.J., and Locke, E.A. (2004). The relationship of achievement motivation to entrepreneurial behavior: a meta-analysis. *Human Performance*, 17, 95–117.

Costa, P.T. Jr, and McCrae, R.R. (1992). *NEO PI-R professional manual*. Odessa, FL: Psychological Assessment Resources, Inc.

Diefendorff, J.M., and Chandler, M.M. (2010). Motivating employees. In S. Zedeck (ed.), *Handbook of Industrial Organizational Psychology*, Vol. 3, pp. 65–136. Washington, DC: American Psychological Association.

Diener, E., Ng, W., Harter, J., and Arora, R. (2010). Wealth and happiness across the world: material prosperity predicts life evaluation, whereas psychosocial prosperity predicts positive feelings. *Journal of Personality and Social Psychology*, 99, 52–61.

Fried, Y., and Jullerat, T. (2011). Work matters: Job design in classic and contemporary perspectives. In S. Zedeck (ed.), *Handbook of Industrial and Organizational Psychology*.

Frye, G.D.J., and Lord, C.G. (2009). Effects of time frame on the relationship between source monitoring errors and attitude change. *Social Cognition*, 27, 867–882.

Locke, E.A., and Baum, J.R. (2007). Entrepreneurial motivation. In

R.J. Baum, M. Frese, and R.A. Baron (eds), *The Psychology of Entrepreneurship*, pp. 93–112. Mahwah, NJ: Lawrence Erlbaum.

Locke, E.A., and Latham, G.P. (2002). Building a practically useful theory of goal setting and task motivation. A 35 year odyssey. *American Psychologist*, 57, 705–717.

Markman, G.D., Baron, R.A., and Balkin, D.B. (2003). The role of regretful thinking, perseverance, and self-efficacy in venture formation. In J. Katz and D. Shepherd (eds), *Advances In Entrepreneurship, Firm Emergence, and Growth*, pp. 73–104. Elsevier/JAI: New York.

Markman, G.D., Baron R.A., and Balkin, D.B. (2005). Are perseverance and self-efficacy costless? Assessing entrepreneurs' regretful thinking. *Journal of Organizational Behavior*, 26(1), 1–19.

Owen, D. (2012). The artificial leaf: Daniel Nocera's vision for sustainable energy. *The New Yorker*, 14 May, 68–74.

Rindova, V., Barry, D., and Ketchen, D.J., Jr (2009). Entrepreneuring as emancipation. *Academy of Management Review*, 34, 477–491.

Shane, S. (2008). *The Illusions of Entrepreneurship: The costly myths that entrepreneurs, investors, and policy makers live by*. New London, CT: Yale University Press.

Tung, R.L., Walls, H.J., and Frese, M. (2006). Cross-cultural entrepreneurship: The case of China. In R. Baum, M. Frese, and R.A. Baron (eds), *The Psychology of Entrepreneurship*, pp. 265–286. Mahwah, NJ: Erlbaum.

Webb, T., and Sheeran, P. (2007). How do implementation intentions promote goal attainment: A test of component processes. *Journal of Experimental Social Psychology*, 43, 295–302.

Zhao, H., Seibert, S.E., and Hills, G.E. (2005). The mediating role of self-efficacy in the development of entrepreneurial intentions. *Journal of Applied Psychology*, 90, 1265–1272.

Zhao, H., Seibert, S.E., and Lumpkin, G.T. (2010). The relationship of personality to entrepreneurial intentions and performance: A meta-analytic review. *Journal of Management*, 36, 381–404.

3 Human cognition: the internal origins of creativity, innovation, and ideas for the *possible*

Chapter outline

Human cognition: impressive, but imperfect
- Memory: how we store, recover, and use information gained through experience
- Cognitive errors and bias: why, often, we are far from rational
  - The confirmation bias
  - Beyond the confirmation bias: the powerful effect of wishful thinking
  - Fast thinking effect: why risk-taking increases when the clock is ticking
  - Heuristics: quick-and-simple—but inaccurate—rules for making judgments and decisions
  - The self-serving bias
  - Optimistic bias and planning fallacy
  - Affect infusion: how feelings shape thought
  - Sunk costs: getting trapped in bad decisions
  - The last is best effect
  - How can entrepreneurs reduce the impact of cognitive biases?

Creativity: where entrepreneurship begins
- Concepts: what they are, and why they are definitely a "two-edged sword"
- Creativity: its foundations in the unusual or obscure

Opportunity recognition: from ideas to action
- Cognitive foundations of opportunity recognition: alertness and connecting the dots
- Alertness: three basic components
- Pattern recognition: connecting the dots to recognize opportunities

* * *

> Creativity can solve almost any problem. The creative act, the defeat of habit by originality, overcomes everything.
>
> (George Lois)

> Creativity is . . . seeing something that doesn't exist already and finding out how you can bring it into being.
> (Michele Shea)

We live in what has often been termed "the digital age": many people—especially those below the age of 30—seem to exist on-line as much as they do in the physical world. They spend hours on Facebook, Twitter, email, and many other digital sites and activities. One result of this trend seems to be that an increasing number of persons—again, primarily young ones—are lacking in basic social skills. They *want* a real social life as well as an electronic one, but are deficient in the social skills that function in face-to-face situations to build connections between people, especially those meeting for the first time. As a result, such individuals often feel isolated and unable to develop the real social relationships that they desire, no matter how many digital "friends" they have.

That is the problem; what is the solution? One might be to somehow persuade these persons to spend less time at on-line sites, but that seems to be a very tall order. Another—and one that is, perhaps, a bit more unusual and creative—is to provide people who feel that they are lacking in social skills with someone who has these skills, and can use them for their (i.e., the customer's) benefit. In fact, several entrepreneurs have recognized this opportunity, and developed companies to provide precisely this kind of service. They offer socially-skilled individuals-for-hire, people who can accompany less skilled individuals to social gatherings, where they help them "break the ice" with persons they want to meet—and then, at precisely the right moment, fade away. That is, they hire a "wingman" or "wingwoman" to help them. Here, slightly paraphrased, is what one such company promises on its website: "The person you hire will:

- "Show you where to party;
- Help you approach a woman/man of your interest and break the ice;
- Save the conversation when it goes bad;
- Disappear at the right moment;
- Reduce the risk of your being rejected by a woman/man".

Does this kind of service work? It is too early to say, but these companies are doing a rising rate of business (see Figure 3.1).

Why do we begin with this new kind of business? Because it provides a clear illustration of the importance of two basic cognitive processes in entrepreneurship: (1) creativity—which basically involves coming up with an idea for something new, useful, and better, and (2) opportunity recognition or identification—devising ways to turn these ideas into profitable business ventures.

Note: Perhaps because they spend so much time "on-line", many young people are relatively low in social skills. One solution, identified by entrepreneurs: provide them with a *wingman* or *wingwoman*, someone who *is* socially skilled to help them "break the ice".

Source: Fotolia 29296300.

Figure 3.1 *Making social contacts—with the help of a wingman or wingwoman*

In the present chapter, we will focus on cognition—the internal (mental) processes that help us understand the world around us, and adapt effectively to it. As we will soon see, these processes truly provide a key foundation for entrepreneurial activities—and entrepreneurial excellence. Entrepreneurship proceeds, in large measure from the creativity of individual entrepreneurs, and its ultimate outcomes are strongly determined by how well they perform a wide range of cognitive tasks: attention (staying focused, recognizing what's import and what's not), memory (retrieving and using information obtained in the past), solving problems, making decisions, and many other activities having to do with the acquisition, processing, and use of information (e.g., Baron, 2007). To examine the essential role of cognition in entrepreneurship and a tool for entrepreneurial success, we will proceed as follows. First, we will examine some of the basics of cognition, especially memory, and some of the errors and biases to which it is prone. Next, we will turn to *creativity*, examining its foundations, its role in innovation, and how it can be enhanced. Third, we will examine *opportunity recognition*—a central process in entrepreneurship, and one in which the products of creativity are used to generate ideas that form the basis of new ventures or other expressions of the entrepreneurial spirit.

Human cognition: impressive, but imperfect

Where were you when the tragic events of 9/11 occurred? Can you recall your first day of school or how your mother and father looked at that time? Who was your first love, and what did it feel like to be in love with this person? Can you read or speak more than one language? Dance? Play golf or tennis? Do you intend to start a company in the foreseeable future? If so, what are your ultimate goals for it? All of these activities derive from, and involve, our basic cognitive systems. Without them, we could not make decisions, remember information, reason, plan, talk, think, or perform any skilled activity. Clearly, then, our cognitive abilities are crucial for almost everything we do, think, or feel. They are also the source of ideas, our creativity, goals, and plans for reaching them—in short, for central aspects of entrepreneurship. Research on human cognition has continued for decades, and uncovered an amazing array of facts about it—especially recently, with the advent and use of techniques for measuring brain activity and function (e.g., Amodio, 2010). Here, though, we will focus on two topics that are highly relevant to entrepreneurial excellence: memory, and a vast array of cognitive errors and biases (e.g., Ariely, 2009).

Memory: how we store, recover, and use information gained through experience

Some people suffer from rare disorders in which they cannot retain information for more than a few seconds. For such individuals, each day is like the start of a new existence: they cannot recall anything that happened to them previously. In most cases, they can remember how to speak, get dressed, and perform many different tasks. But in other respects, their minds are like blank canvases, waiting to be filled again each day.

Fortunately, very few persons face this kind of situation. Almost all of us possess functioning *memories*—cognitive systems that allow us to store, retrieve, and use information acquired previously. As you can guess, these systems are crucial for many important tasks, from solving problems to making plans, and underlie creativity, which often involves combining information in new and unexpected ways (e.g., Sternberg and Sternberg, 2011). Actually, we possess three distinct kinds of memory. One, *working memory*, holds a limited amount of information for brief periods of time—perhaps up to a few seconds. If you look up a phone number and then try to remember it just long enough to dial it, you are using working memory. In a sense, working memory is where our current consciousness exists: it is the system that holds information we are processing or using now, so in a sense, it is where our consciousness exists.

Another, and very different system, is known as *long-term memory*.

This system allows us to retain truly vast amounts of information for long periods of time. Research findings indicate that there are no clear limits to how much information long-term memory can hold or how long it can be retained. That is one basic reason why we can continue to acquire new knowledge and new skills throughout life. For instance, my wife's father learned to play the tuba when he was in his 70s, after retiring from a long and happy career as a physics professor. Long-term memory, like other biological systems, does decline somewhat with age, but the rate is very slow and barring serious illness, memory can function very well throughout the lifespan.

Long-term memory can hold many kinds of information—factual knowledge (e.g., what is the price of a competitor's product, or the current share price of Facebook?). It also contains personal knowledge about events we have experienced in our own lives (e.g., what are known as *autobiographical memories*).

Another kind of information retained in long-term memory is much harder to express in words. For instance, a champion chess player cannot easily explain how she or he anticipates opponent's moves, and a champion golfer cannot explain how he or she hits the ball so far or so straight. Interestingly, experienced entrepreneurs often remark that they can distinguish a potentially valuable opportunity from one that is probably unlikely to succeed very quickly— almost instantly—and, again, they cannot describe clearly how they do this. As one put it to me: "I don't know how I know—I just *do*." This capacity is related to expertise (Baron and Henry, 2010), because one effect of developing expertise in almost any field is that it provides the persons involved with amazingly rapid access to information stored in long-term memory. Remember: that system can hold vast amount of information, so a key problem is finding information in it. Being able to do that very quickly—and accurately—may be one of the reasons why experts in a given field perform at such amazingly high levels.

This kind of memory—memory for information that cannot be readily put into words—is known as procedural memory. It is a kind of long-term memory because information stored in it can be retrieved years, or even decades later. For example, if you learn to swim as a child you can probably do so many years later, even if you have had no opportunity to swim since.

It is both important and interesting to note that our decisions, goals, and plans are often strongly influenced by information stored in procedural memory or in other aspects of memory that are outside conscious awareness. We cannot describe such information and often are only dimly aware of its presence, but it influences important aspects of our cognition nevertheless. Such information is the basis for intuition, which is sometimes described as "off-line processing" because it involves processing of information that occurs outside our normal stream of conscious experience. It is also reflected in tacit knowledge, knowledge that involves what is often described as "know-how"—knowing how to perform various actions (e.g., complex mathematical computations, speaking another language) rather than "know-what"—knowledge of specific information. Because tacit knowledge is represented in procedural memory, it is often difficult to share with others, who must learn from observing what we do, rather than from our descriptions of these actions and how we perform them.

Cognitive errors and bias: why, often, we are far from rational

That our cognitive abilities are impressive is obvious. We can often retain information for many years, recognize thousands of different persons, and (some of us, at least!) solve tremendously complex equations, speak many languages, or generate beautiful art or music most of us can only appreciate, not create. Impressive as our cognitive abilities are, however, they are far from perfect. In fact, they are prone to a wide range of errors—ones that can badly distort our judgment, decisions, and reasoning (Ariely, 2009). You already know from your own experience that memory is definitely flawed: on many occasions, we cannot retrieve information from it, often, just when we need it most. And memory is frequently inaccurate, so that when information is found, it is not what we want, or has been changed in various ways. In fact, evidence indicates that each time information is retrieved from memory and then re-entered, it tends to shift—often, so that it becomes more consistent or easier to interpret. For instance, I have been asked about my own experiences as an entrepreneur so many times, that now, when I try to recall these events, the information I retrieve from memory is almost certainly more flattering to me and my abilities than is probably true. But errors in memory are only the tip of the iceberg, so we will now examine several others that can, unfortunately, drastically interfere with entrepreneurial excellence. For that reason alone, it is important for you to know about their existence. Here are some of the most intriguing—and most important.

The confirmation bias

If we were totally rational processors of information, we would be especially sensitive to information that is inconsistent with our current beliefs or views. Such information would be very helpful in terms of allowing us to examine these beliefs, and to determine if they are in fact accurate. Unfortunately, we show precisely the opposite tendency. Most people, most of the time, have a strong preference to notice, process, and store only information that confirms their existing beliefs; they filter out or discount the rest, and do not let it remain in working memory and then pass through to long-term memory. As a result, they become locked into what have been termed "inferential prisons"—external information that is inconsistent with their current thinking tends to be ignored rather than change it. Clearly, this is something entrepreneurs—who must pay careful attention to new information and conditions—should avoid.

Beyond the confirmation bias: the powerful effect of wishful thinking

Suppose an entrepreneur strongly believes that the product she plans to develop is far superior to existing, competing products. Certainly she wants to believe that this is true. But now, imagine that the entrepreneur receives information from two separate marketing studies. One indicates that in fact, consumers do find the new product superior to existing ones. In contrast, the other study indicates exactly the opposite: consumers find the current products to be superior to the new one. The confirmation bias predicts that the entrepreneur will tend to believe the survey favorable to her or his product. But will she also perceive the *quality* of the survey that contradicts her beliefs to be lower than that of the one that confirms these views? Research findings indicate that she will. In other words, her evaluation of the two studies will be strongly affected by wishful thinking—what she *wishes* to be true. Again, this is a tendency entrepreneurs should recognize—and resist.

Fast thinking effect: why risk-taking increases when the clock is ticking

It has often been assumed that entrepreneurs are risk-takers: they are more willing to accept high levels of risk than other persons. Although evidence on this suggestion is mixed (in fact, some indicates that entrepreneurs are less accepting of risk than others; e.g., Miner and Raju, 2004), recent findings indicate that a basic aspect of our cognitive systems may tend to encourage risk-taking among entrepreneurs. Specifically, it has been found that people think quickly—as, for example, when they face deadlines, or in situations that require quick responses—their tendency to accept high levels of risk increase (Chandler and Pronin, 2012). Why? Perhaps because fast thinking signals a need for immediate action, and this encourages daring behavior. Fast thinking may also tend to elevate mood, and more positive moods may increase confidence, and so reduce the perceived levels of risk present in a situation (e.g., Baron et al., 2012). Whatever the basis, this is certainly a cognitive bias with potentially dangerous effects for entrepreneurs.

Heuristics: quick-and-simple—but inaccurate—rules for making judgments and decisions

Often, we encounter more information than we can possibly process at a given time. This leads to a strong tendency to use heuristics—quick (and sometimes "dirty") rules for making complex judgments and decisions. Several exist, but two that are especially powerful should be briefly mentioned. The first, known as the availability heuristic or rule, suggests that whatever we can bring to mind most easily tends to be viewed as most important or accurate. So, if we can remember information easily, it has a powerful impact on our current thinking. Unfortunately, such information is sometimes easy to recall because it is vivid or unusual, not because it is particularly useful or revealing.

Another heuristic—and one that is especially important to entrepreneurs—is the anchoring-and-adjustment heuristic. This refers to our powerful tendency to accept an opening price or position as an anchor from which adjustments can be made. In fact, such initial offers or positions may have little relationship to reality; yet, they strongly influence our judgments. This is why retailors try to set the price at an ideal point: high enough to influence buyers to pay more, but not so high that they find the initial price unreasonable and walk away. If you have ever watched the popular television show *Pawn Stars* you have seen the anchor-and-adjustment heuristic in operation. In this show, the owners of a large pawn shop attempt to negotiate the price of various items people bring to them to sell. Usually, the owners ask the would-be sellers to name a starting price—mainly so they can indicate, immediately, that this is completely unreasonable. (The phrase one of them uses says it all very succinctly: "No way that's going to happen"—meaning that there is no way he will pay the asking price.) After this, they name a price of their own, and the sellers—who are far less experienced as negotiators—usually adopt this an anchor and then try to make small adjustments to it (in their favor of course). The outcome is almost always the same: the pawn shop owners (the experts) buy the items for a very favorable price. The moral: beware of the anchor-and-adjustment heuristic, because it can prove very costly in negotiations—a process of major importance for entrepreneurs.

The self-serving bias

Another strong tendency in our thinking is to attribute favorable outcomes to our own effort or talent, while negative outcomes are attributed to external factors beyond our control. The result is that individuals do not learn from their errors—because they do not see them as errors; instead, they perceive negative outcomes as "not their fault", and so discount them instead of examining them very carefully. This bias can also lead to considerable friction between partners (e.g., founders of a new company), since each assumes that *they* are primarily responsible for positive outcomes, while their partners are responsible for negative ones.

Optimistic bias and planning fallacy

People are, by and large, very optimistic: they tend to believe that things will turn out well, even if there is no rational basis for such beliefs. Closely related is the tendency to underestimate the amount of time needed to complete a given task, or to assume that more can be completed in a given period of time than is feasible. Together, these cognitive biases can lead to serious errors in planning, formation of strategy, and many other activities. Moreover, growing evidence indicates that they are truly a basic aspect of human thought—tendencies "built in" to the structure and function of our minds and brains (e.g., Sharot, 2011). Recent research indicates that the optimistic bias is directly relevant to entrepreneurship and in ways that seem to contradict widespread beliefs: the more optimistic entrepreneurs are, the poorer the performance of their companies (Hmieleski and Baron, 2009). Why? Apparently because high levels of optimism interfere with setting appropriate goals and with effective processing of relevant information, thus interfering with effective decision-making.

Affect infusion: how feelings shape thought

If we were totally rational beings, our feelings and emotions would not influence our decision or judgments. Do they? Of course! A large body of research evidence indicates that feelings, and even mild and subtle shifts in moods, often exert powerful effects on our thinking. For instance, when we are in a good mood, we tend to recall mostly positive information; when we are in a bad one, we tend to remember mostly negative information. Clearly, what we recall can strongly influence our decisions, so affect infusion—the tendency for our current feelings to influence key aspects of our cognition—is another potentially important source of cognitive errors. (We will consider the role of emotions and feelings in entrepreneurship in detail in Chapter 4, where we discuss such topics as entrepreneurial passion and the question of whether entrepreneurs can sometimes be too positive or enthusiastic.)

Sunk costs: getting trapped in bad decisions

Have you ever repaired a used car, only to have it break down again very soon. What do you do then? Rationally, at some point, you should "walk away" and write-off the money you have already spent on the car as wasted. But in fact, most people find it very difficult to do this. Instead, they feel psychologically committed to their previous decisions, and cannot walk away. They get trapped in what are called "sunk costs"—the resources they have already invested in a failing course of action. This is a very powerful tendency, and can badly distort the thinking and actions of individuals who become its victim.

The last is best effect

J.K. Rowling wrote seven "Harry Potter" stories, completing the last one in 2007 (*Harry Potter and the Deathly Hallows*). If readers of these books were asked which the best was, would there be a clear winner? Growing evidence indicates that the last would probably be viewed as the best (e.g., O'Brien and Ellsworth, 2012), even if, by objective standards, this was not true. Why? Perhaps because they know that no more will be coming, or because knowing that something is the "last" shifts attention to positive aspects of the experience. Whatever the reason, this effect, too, has important implications for entrepreneurs. In business plan competitions, entrepreneurs give short descriptions of their companies to a panel of judges, who then select the ones they view as best; these potential new ventures receive large cash prizes, and are also introduced to venture capitalists and others who may provide even more start-up funding. The "last is best" effect suggests that there may be an important advantage to being the last group to present. Given the intensity of such competitions, that's an important piece of information for entrepreneurs!

Many other cognitive errors and biases exist as well but by now, the main point should be clear. Partly because of the limitations of our own cognitive systems, we are far from totally rational as information processors. On the contrary, we are susceptible to many errors and biases which, together, often interfere with our capacity to make accurate and effective decisions, judgments, and choices. As shown in Table 3.1, and as noted throughout this discussion, it is important for entrepreneurs to be familiar with them, and to guard against their impact. Doing so can add significantly to their entrepreneurial excellence in several different ways.

How can entrepreneurs reduce the impact of cognitive biases?

If these "tilts" in our cognitive systems are general—ones experienced by most persons—how can they be reduced? Fortunately, several techniques for reaching this goal exist. Often, simply knowing about their existence can be very helpful. For instance, Ariely (2009, p. 244) suggests that once we understand our susceptibility to such errors, and their basic nature, we can reduce or avoid them by being vigilant and actively trying to think in ways that minimize their impact. Technology, too, can help; for instance, it can remind us—vividly!—of the strengths of competitors, and the odds of failure for new products in highly competitive markets. This can help to reduce our strong optimistic bias. As another researcher puts it (Sharot, 2011): "Once we are made aware of our optimistic illusions, we can act to protect ourselves." In essence, enhanced self-knowledge may be one basis for mitigating the impact of our built-in tendencies to fall prey to cognitive

Table 3.1 *Cognitive errors that can be especially costly for entrepreneurs*

Cognitive Error	*Description*	*Relevance for entrepreneurship*
Confirmation Bias	Tendency to notice, process, and store only information consistent with current beliefs	Reduces capacity to be flexible in the face of changing conditions, and capacity to respond to negative information
Heuristics	Rules of thumb for making decisions and judgments quickly	Efficient in terms of reducing cognitive effort, but can lead to serious errors when more systematic and detailed analysis is required
Self-serving Bias	Tendency to attribute positive outcomes to one's own talent, effort, etc., but negative ones to external factors beyond one's control	Reduces capacity to learn, since negative outcomes are perceived as generated by external agencies or factors
Optimistic Bias	Tendency to expect more positive outcomes than is rationally justified	Leads to unrealistically high goals and aspirations, and to underestimating the amount of time or effort needed to complete various tasks
Fast Thinking Effect	Faster speeds of thinking enhance risk-tasking	Entrepreneurs must often make decisions rapidly and this may increase their tendency to assume high risks
Affect Infusion	Influence emotions and feelings on key aspects of cognition (e.g., decision-making, evaluation of various alternatives)	Can seriously distort judgments and decisions by entrepreneurs in a wide range of contexts
Sunk Costs	Tendency to get trapped in bad decisions or failing course of action	Can prevent entrepreneurs from "cutting their losses" by walking away from poor decisions or strategies
The last is best effect	The last in a series of events is perceived most positively	Entrepreneurs who present last in business plan competitions or other contexts may gain an advantage

Note: All of the errors listed here can interfere with entrepreneurs' capacities to make accurate decisions and judgments, as well as other important aspects of cognition.

errors. Only future research will reveal whether, and to what extent, that is the case, but it is consistent with other evidence suggesting that "to be forewarned is, often, to be forearmed" (e.g., Baron and Branscombe, 2012). In short, the fact that you are now familiar with some of the most important forms of cognitive errors may help you to resist them. So please: do carefully consider the information in this discussion, and the summary in Table 3.1. This may well help you to avoid telling yourself "I should have known better" when, looking back, you consider your own mistakes and bad decisions at later times!

Creativity: where entrepreneurship begins

Entrepreneurship is clearly a process—not a single event (e.g., Baron and Shane, 2008). It starts with an idea for something that will solve an existing problem or meet an existing need in ways that are better than current means for doing so, proceeds with efforts to determine if this idea might provide the basis for entrepreneurial activity (e.g., starting a new venture, developing something new within an existing company), turns to development of plans to actually develop this perceived opportunity, and then on to actions that turn these possibilities into realities. For this reason, it is difficult to point to a particular moment as the start of the entire process. Yet, that said, there are strong grounds for suggesting that the entire process begins with creativity—the emergence of an idea for, or image of, something that is both new and useful. As we will note below, measuring creativity is a difficult and complex task, but often, it is like many other complex concepts—difficult to define precisely, but easy to recognize. In other words, we know it when we see it! Here's an example from my own life.

About twelve years ago, I was visiting my brother and saw him use a kitchen tool I had never seen before: a new kind of grater. He used it to quickly grate the skin of several lemons—a task that often results in raw fingers and torn fingernails, and then used it to grate some spices. My first thought was: "How useful!" followed by a question: "Where can I get one!" What my brother was using was a Microplane, a product that is now standard equipment for leading chefs, and many millions of cooks, around the world. In fact, it is so superior to what existed before, that it has quickly replaced earlier types of graters, and has recently—despite strong patent protection—generated a host of imitators. How did the idea for it emerge? As we will see below, it derived from a kind of file used by woodworkers so like almost all "new" ideas, it did not spring from a cognitive vacuum: rather, it developed out of modifications in existing information and knowledge.

One important point before continuing: even if a creative idea is recognized as being both new and potentially useful, this is no way assures that it will be accepted. In fact, recent evidence indicates that there is a subtle, implicit bias against creativity—or at least, the "new" portion of it. Why? Because people dislike uncertainty, and anything new generally increases it. So people may overtly express approval and support for creativity, but implicitly reject it, or least be wary of it (Mueller et al., 2012).

The Microplane also provides a good example of the link between creativity and innovation. Innovation takes the process one step farther, by applying a creative idea to something concrete that is (1) better than what existed before, and (2) can be used to generate value (economic or social). That, as we hope you will recall, is the essence of entrepreneurship. As Amabile (1996, p. 143), a leading expert on creativity puts it: "All innovation begins with creative ideas . . . creativity by individuals and teams is a starting point for innovation." In other words, the products of creativity—new ideas, principles, or concepts—serve as the "raw materials" for innovation. Thus, creativity is often a necessary condition for subsequent innovations, although not a sufficient one, since many ideas generated by creativity are not commercially feasible or cannot be developed by the persons who generated them (e.g., McMullen and Shepherd, 2006; Ward, 2004). Research findings indicate that creativity does indeed enhance innovations by new ventures (Baron and Tang, 2011), although this relationship is stronger in highly dynamic environments (i.e., ones that are rapidly changing) than in more stable environments.

It is interesting to note that the United States Patent Office applies very similar criteria to patent applications: to qualify for a patent, an idea must be new, not obvious to anyone familiar with similar ideas or products that already exist, and useful, in the sense that it can actually be produced or put to use. In short, creativity is truly a building block of entrepreneurship. But how does this process actually unfold? How, and why, do specific individuals come up with ideas for something new that is also useful? Since creativity is clearly a result of human cognition—ideas originate in the minds of specific persons—it is useful, once again, to consider what basic knowledge of cognition tells us.

Concepts: what they are, and why they are definitely a "two-edged sword"

As we noted earlier, information entered into long-term memory has a nasty habit of getting lost—like a missing computer file, we cannot find it when we need it. To reduce this problem, information in memory is organized in various ways—it is grouped together in internal frameworks that enhance its ease of recall. Several types of mental frameworks exist and are useful, but the ones most relevant to creativity are known as concepts. Concepts are a kind of mental "container" with objects, events, or ideas that are somehow similar to each other. For instance, the words automobile, airplane, and boat are all included in the concept *vehicle*, because all share certain key features: they move people around, have controls to start and stop them, and to change direction, and so on. Similarly, consider the words shirt, jeans, hoodies, and shoes. All fit within the concept *clothing*, because they share several features: they are worn by people, can be put on or removed, cover various parts of the body, and so on. A key feature of concepts—and one closely related to their role in creativity—is that they all get "fuzzy" around the edges. Consider the concept of vehicle once again.

Does elevator fit? Probably it does—it carries people and can start and stop, but does not ordinarily change direction. But what about escalator? Roller skates? Similarly, think about the concept clothing. Shirts, jeans, and shoes are at the core of this concept; but what about a wig? And what about tattoos—they cover parts of the body, but do not fit well within the concept clothing.

The fact that concepts get "fuzzy" at the edges is closely related to their role in creativity. Briefly, concepts can encourage creativity when they are expanded or combined in various ways. The fact that they get "fuzzy" at the edges can greatly facilitate these processes.

First, let us see how concepts can be combined to generate something very new. Do you use a "smart phone"? If so, you know that such phones combine many concepts into a single new product: they are phones, GPS devices, cameras, music systems, tools for finding the prices of various products in nearby stores, for gaining access to the Internet, and many other purposes—the list goes on and on and is limited only by the number of "apps" owners obtain and use. Smart phones represent a combination of several concepts, and this combination is so useful to many people, that they can no longer imagine life without these phones. In instances like this, creativity is greatly enhanced through the combination of several initially separate concepts.

Another route to the same goal is concept expansion. This happens when existing concepts are mentally stretched so that they generate ideas for new applications. For instance, the Microplane, described above, developed out of the existing concept for "rasp"—a kind of file used to produce a smooth finish on wood. This concept is ancient—rasps were used in ancient Egypt, 5000 years ago. But the inventor of the Microplane, a skilled woodworker, stretched this concept so that its basic features (a hand-held tool containing many small teeth that remove material from the surface of objects) were stretched and adapted for use with various foods—lemons, cheese, spices, and so on (see Figure 3.2).

Clearly, then, the existence of mental "bins" or containers in our minds—concepts—can be a major "plus" in terms of enhanced creativity. But as noted briefly above in connection with the Inca's failure to use the wheel to transport various items, there is an important "downside" to this story. If they are very strong and clear, concepts can lock us into traditional ways of thinking—what I personally describe as "mental ruts". Unfortunately, these "mental ruts" can be very deep so that people find it very difficult to escape from them. Here is one intriguing example. In the 1960s, engineers at Sony Inc. were given the task of developing a new means for storing music. Working with the technology available at the time, they came up with an early form of the CD. It worked, but was rejected as a potential product for one simple reason: it was far too large

The Microplane

A rasp for smoothing wood

Note: The Microplane, a very popular new product, was derived from the basic concept for a rasp—a tool used to smooth wood since ancient times. By expanding this concept, the basic idea for a new and highly useful product was produced.

Sources: Robert A. Baron; Fotolia 15669504.

Figure 3.2 *Expansion of concepts—an example*

to be portable. Why did the engineers make this mistake? Because they were trapped, by their own experience with large long-playing records, into assuming that this new storage device, too, had to be twelve inches in diameter (see Figure 3.3)! In sum, concepts can either encourage or obstruct human creativity; the key task is to use them as a basis for creative ideas, while avoiding the mental ruts into which they sometimes direct our thinking.

Creativity: its foundations in the unusual and obscure

As described above, concepts exert powerful effects, and sometimes, it is extremely difficult to break free of their influence in order to think creatively. Growing evidence indicates, however, that successful escape from "mental ruts" involves attention to what might be termed the unusual and the obscure.

In this context, the term *unusual* refers to avoiding high-frequency, typical associations, and recognizing, instead, ones that are much less likely to occur. For instance, one popular test of creativity is the Remote Associates Test (e.g., Topolinski and Struck, 2009). This test involves showing persons taking the test three seemingly unrelated words (e.g.,

Note: When Sony engineers were given the task of developing a new means for storing music, they came up with a device that was 12 inches in diameter—the same size as the records it was designed to replace (left photo). There was no technical reason why the new devices—early forms of the CD—had to be this large (right photo). The engineers were trapped by their own experience and concepts, into assuming that the new devices had to be so large!

Source: Fotolia 4218005 left hand, Fotolia 294893 right hand.

Figure 3.3 *How concepts sometimes restrain creativity: the impact of mental ruts*

surprise, line, birthday), and asking them to come up with a fourth word that relates to all of them. In this instance, the correct answer is party. This answer is a high-frequency association for each of the unrelated words—e.g., "party" is a strong association to the word "birthday". Most persons, therefore, can solve this problem. But when high association words are not correct, many individuals have great difficulty in providing the correct answer. Recent studies (e.g., Gupta et al., 2012) indicate that being able to go beyond such high-association answers is an important component of creativity. In other words, creative persons are ones who can ignore the usual or typical, and make associations with the unusual or atypical. For instance, they can provide the answer *panel* to the words, jury, door, and side. In contrast, most persons get stuck on problems like this one, because they can think only of the usual associations to each word—e.g., trial for jury and step for door. So, an important source of creativity may be the capacity to think of the unusual rather than the typical.

Other, and related research, suggests that creativity and innovation may also stem from recognizing the obscure—what is not obvious at first glance. This is shown by research using another test of creativity—what is known as insight problems. In these problems, individuals are given several objects, and asked to figure out a way to use them solve a problem. One such task includes two heavy metal rings, a candle, a match, and a 2-inch cube of steel (McCaffrey, 2011). The task is figuring out some way to attach the two rings together. How would you solve this problem? Most people try to melt the candle and use the wax to join the rings, but this does not work—they are too heavy. The correct answer involves searching for the obscure, which, in turn, involves examining the basic parts of each object, and using these to solve the problem. In this case, the candle consists of wax and a string (the wick). By removing the wax, the string is exposed, and can be used to tie the two rings together. People who perform this kind of analysis on their own—generic parts technique—are better at solving such problems than people who do not, and even brief training in using this approach can greatly increase almost anyone's performance (e.g., McCaffrey, 2012). This suggests that creativity can in fact be increased: it is not something set in stone or in our genes. And in fact, a large body of research (e.g., Sternberg and Sternberg, 2011) offers support for this view. To the extent concepts can be combined or expanded, and individuals can learn to focus on the unusual or obscure—ideas very close to the common phrase "think outside the box"—they can enhance their own creativity. And since creativity is so central to entrepreneurship, doing so can contribute significantly to their entrepreneurial excellence.

Opportunity recognition: from ideas to action

Have you ever suffered from hiccups? If so, you know that sometimes, getting them to stop is very difficult. In fact, for some people, this can be a serious problem: if they experience hiccups while driving or when speaking in front of a group, the consequences can be annoying, embarrassing—or worse. Until recently, there was no sure-fire way of combating hiccups. Breathing into a paper bag, holding one's breath, or being surprised were not very effective. Recently, however, a 13-year-old girl name Mallory came up with a solution—one medical researchers never considered. She has invented the "Hiccupop"—a lollipop that combines apple-cider vinegar and sugar in a lollipop. Licking it overstimulates the nerves in the throat, and blocks the neural message for hiccups. As a result, almost all persons who lick the pop during hiccups obtain relief—the hiccups stop! Mallory has applied for a patent, and is now working with MBA students at the University of Connecticut to bring this invention to market.

Why did she and these students decide to move forward? Because they perceived a major opportunity in this product. In short, they perceived something that was (1) new, (2) useful, (3) could potentially generate profit or other beneficial outcomes, and (4) desirable—it is legal and non-harmful, and does not violate any moral principles or strong social norms. In short, an opportunity, in the context of entrepreneurship, is defined as *perceived means of generating value (i.e., profit or other benefits) that are not currently being exploited and are perceived, in a given society, as desirable or, at least, socially acceptable*.

But how do entrepreneurs recognize opportunities? Clearly this is a cognitive event, but how, precisely, does it occur? Several different views of this process exist (e.g., Gregoire et al., 2010), and all add to our understanding of this key step in entrepreneurship. Here, though, we will focus mainly on two for which there is growing empirical evidence.

Cognitive foundations of opportunity recognition: alertness and connecting the dots

Are some people better at recognizing opportunities for entrepreneurial action than others? Interestingly, although economists are not typically concerned with such questions, it was a well-known economist—Israel Kirzner—who first emphasized this possibility. Kirzner (1979) suggested that some individuals—and some entrepreneurs—are very good at this task: they are able to recognize opportunities that most other persons overlook. Kirzner described such persons as high on a dimension of alertness, and added that since noticing opportunities is a crucial first step to deciding to pursue them, alertness is a key variable in entrepreneurship. On the face of it, this was a very reasonable suggestion. But Kirzner simply called attention to this variable and, being an economist, went no further in trying to understand its origins and nature. Why are some persons more alert to opportunities than others? Kirzner did not address this basic question, but recently, there have been many efforts to answer it by understanding the cognitive foundations of alertness (e.g., Alvarez and Barney, 2007; Gaglio and Katz, 2001; McMullen and Shepherd, 2006).

Alertness: three basic components

Among the most informative work to date is that reported by Tang et al. (2012). After carefully reviewing existing evidence and theory concerning alertness, Tang et al. (2012) proposed that it involves three basic dimensions: (1) *alert scanning and search*—continuing efforts by current or future entrepreneurs to identify opportunities for doing something better; (2) *alert association and connection*—continuing efforts to connect or integrate various sources of information, to perceive links between them and use these as a basis for creating something new and useful; and (3) *evaluation and judgment*—efforts to distinguish between high and low-potential opportunities and being able to choose the best ones. To measure individual differences in these dimensions, Tang et al. (2012) developed a questionnaire and tested it extensively to determine if it does indeed measure the core components of alertness. Results indicated that three clear factors reflecting these dimensions did indeed emerge in responses by a large group of CEOs to the questionnaire. Thus, alertness does indeed seem to consist of these three basic factors.

Additional research was then conducted with large samples of entrepreneurs to see if the three components of alertness were related to other aspects of entrepreneurship. For instance, it was predicted that alertness (all three basic dimensions) would be related to innovation in new ventures and to research and development expenditures in these companies. These predictions were confirmed. Further, it was predicted, and found, that the more "upbeat" entrepreneurs were (i.e., the higher they were in dispositional positive affect), the higher they would score in terms of alertness, since positive affect is related to broadened perceptions of the external world (e.g., Fredrickson and Branigan, 2005). Overall, it appears that the framework offered by Tang et al. (2012) does clarify the basic nature of alertness and confirms its important role in opportunity recognition. As Kirzner (1979) suggested, people do indeed differ in their capacity to recognize opportunities, and this reflects differences in their tendencies to engage in careful, continuous scanning of the environment (an active search for opportunities), their capacity to perceive connections between seemingly unrelated information and events they encounter, and their ability to evaluate the potential of the various opportunities they identify. All of these components can be strengthened with practice and effort, so it seems reasonable to conclude that effort directed to strengthening alertness may be one important step current or would-be entrepreneurs can take to build their own excellence and tip the odds of success in their favor.

Pattern recognition: connecting the dots to recognize opportunities

One of the components of alertness involves a capacity to perceive connections between seemingly unrelated events or trends. This factor takes 'center stage' in another view of opportunity recognition—one that focuses primarily on this process of pattern recognition (e.g., Baron, 2006; Matlin, 2004).

As applied to opportunity recognition, pattern recognition involves instances in which specific individuals notice patterns in the external world that other persons overlook (Baron, 2006). The patterns they perceive then become the basis for identifying new business opportunities. For instance, consider an entrepreneur who noticed the following seemingly unrelated events: (1) the cost of medical procedures has risen sharply in the United States; (2) the number of people needing such procedures, but who cannot afford them, has also risen (despite new legislation designed to assist them); and (3) such procedures are available outside the United States, in excellent hospitals with first-rate physicians. These events might at first seem unrelated, but entrepreneurs have recently connected them and as a result, have recognized a new opportunity: medical tourism. This involves arranging for people in need of medical procedures to travel to countries

Note: Hiccups can be annoying, embarrassing, or even worse, but no good means for stopping them existed until Mallory Kievman invented the "Hiccupop—a lollipop that cures hiccups" (top photo). The young inventor is shown below (second from the left) after ringing the starting bell at the New York stock exchange.

Sources: Photographs courtesy of Adam Kievman and Hiccupops.

Figure 3.4 *One creative business opportunity: a cure for hiccups!*

where it is available at much lower cost (see Figure 3.4). Is this an effective solution to the ever-rising costs of medical treatment in the United States? Only time will tell, but clearly, the idea for this kind of business was generated by cognitive processes in which unrelated trends were "connected" mentally in the minds of several entrepreneurs.

Does pattern recognition actually play a role in opportunity

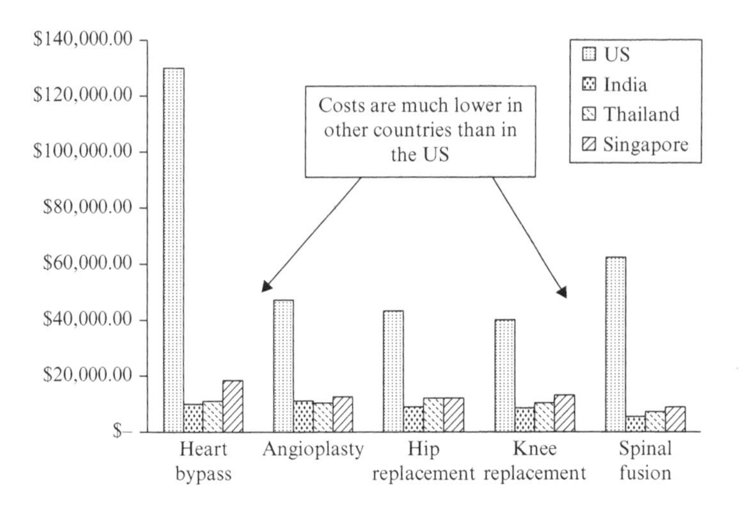

Note: As shown here, the cost of many medical operations is much lower in other countries than in the United States. This fact was combined with several others (e.g., many persons in need of such procedures in the US cannot afford them; excellent hospitals and well-trained physicians are available elsewhere) to generate the idea for a new opportunity.

Source: Based on data from the Medical Tourism Association.

Figure 3.5 *Medical tourism: an opportunity based on pattern recognition*

recognition? Empirical evidence (e.g., Baron and Ensley, 2006) indicates that it does. In fact, one thing that experienced entrepreneurs seem to learn from starting several different ventures and from considering many different potential opportunities, is how to recognize patterns in such diverse events as changes in technology, shifts in government policies and regulations, and demographic trends, such as the rising average age of a population, or shifts in the groups that make it up. Again, skill in pattern recognition can be learned, and by combining such skill with the other components of alertness, entrepreneurs can improve their performance in recognizing potentially valuable opportunities. And since starting with a strong opportunity influences the success of new ventures and other entrepreneurial activities, the overall moral is clear: by learning to look in the right places and in the right ways, current or future entrepreneurs can greatly enhance this important aspect of their own entrepreneurial excellence.

Summary of key points

Human cognition is truly impressive, but far from perfect. Memory, for example, is flawed in many ways. Although we can store seemingly limitless amounts of information in it, finding the information we want is often difficult. In addition, each time information is retrieved from memory and returned to it, the information is changed in subtle ways. In addition, we are subject to a wide array of cognitive errors and biases, including the confirmation bias, self-serving bias, optimistic bias, and many others.

Creativity involves ideas for something new and useful. It rests, in part, on concepts—mental "containers" for ideas, objects, or events that are similar in certain respects. Creativity often results from combining or stretching concepts. In addition, recent evidence indicates that creativity involves attention to what is unusual or obscure. Opportunity recognition is an important initial step in the entrepreneurial process, and involves identifying means for generating value (financial or social) that are not currently being developed. Opportunity recognition rests on important cognitive foundations, including alertness (an active search for opportunities coupled with evaluation) and pattern recognition—noticing potential connections between seemingly unrelated events, trends, or processes. In closing we should note that recent research (Gregoire and Shepherd, 2012) also indicates that opportunity recognition involves aspects of the opportunities themselves—for instance, their similarity to current market needs. In short, opportunity recognition is based on a complex convergence of the actions and thoughts of individuals, *and* the characteristics of the opportunities.

References

Alvarez, S.A. and Barney, J.B. (2007). Discovery and creation: Alternative theories of entrepreneurial action. *Strategic Entrepreneurship Journal*, 1, 11–26.

Amabile, T.M. (1996). *Creativity in Context*. Boulder, CO: Westview Press.

Amodio, D.M. (2010). Coordinated roles of motivation and perception in the regulation of intergroup responses: Frontal cortical asymmetry effects on the P2 event-related potential and behavior. *Journal of Cognitive Neuroscience*, 22, 2609–2617.

Ariely, D. (2009). *Predictably Irrational: The hidden forces that shape our decisions*. New York: Harper Collins.

Baron, R.A. (2006). Opportunity recognition as pattern recognition: How entrepreneurs "connect the dots" to identify new business opportunities. *Academy of Management Perspectives*, 20, 104–119.

Baron, R.A. (2007). Behavioral and cognitive factors in entrepreneurship: entrepreneurs as the active element in new venture creation. *Strategic Entrepreneurship Journal*, 1, 167–182.

Baron, R.A., and Branscombe, N.R. (2012). *Social Psychology*, 13th edn. Boston: Pearson Education.

Baron, R.A., and Ensley, M.D. (2006). Opportunity recognition as the detection of meaningful patterns: Evidence from comparisons of novice and experienced entrepreneurs. *Management Science*, 52, 1331–1344.

Baron, R.A., and Henry, R.A. (2010). How entrepreneurs acquire the capacity to excel: Insights from basic research on expert performance. *Strategic Entrepreneurship Journal*, 4, 49–65.

Baron, R.A., and Shane, S.A. (2008). *Entrepreneurship: A process perspective*, 2nd edn. Cincinnati: Thompson-Southwestern.

Baron, R.A., and Tang, J. (2011). Positive affect, creativity, and innovation in new ventures: A moderated mediation model. *Journal of Business Venturing*, 26, 49–60.

Baron, R.A., Hmieleski, K.M., and Henry, R.A. (2012). Entrepreneurs' dispositional positive affect: The potential benefits—and potential costs—of being "up". *Journal of Business Venturing*. 27, 310–324.

Bastardi, A., Uhlmann, E.L., and Russ, L. (2012). Wishful thinking: Belief, desire, and the motivated evaluation of scientific evidence. *Psychological Science*, 23, 731–372.

Chandler, J.J. and Pronin, E. (2012). Fast thought speed induces risk taking. *Psychological Science*, 23, 370–374.

Fredrickson, B.L. and Branigan, C.A. (2005). Positive emotions broaden the scope of attention and thought-action repertoires. *Cognition and Emotion*, 19, 313–332.

Gaglio, C.M. and Katz, J. (2001). The psychological basis of opportunity identification: The psychological basis of opportunity identification: Entrepreneurial alertness. *Small Business Economics*, 16, 95–111.

Gregoire, D.A., Barr, P.S., and Shepherd, D.A. (2010). Cognitive processes of opportunity recognition: The role of structural alignment. *Organizational Science*, 21, 413–431.

Gregoire, D.A., and Shepherd, D.A. (2012). Technology-market combinations and the identification of entrepreneurial opportunities: An investigation of the opportunity-individual nexus. *Academy of Management Journal*, 55, 753–785.

Gupta, N., Jang, Y., Mednick, S., and Huber, D.E. (2012). The road not taken: Creative solutions require avoidance of high-frequency responses. *Psychological Science*, 23, 288–294.

Hmieleski, K., and Baron, R.A. (2009). Entrepreneurs' optimism and new venture performance: A social cognitive perspective. *Academy of Management Journal*, 52, 473–488.

Kirzner, I. (1979). Entrepreneurial alertness. *Small Business Economics*, 16, 95–111.

Matlin, M.W. (2004). *Cognition*, 6th edn. Fort Worth, TX: Harcourt College Publishers.
McCaffery, T. (2011). The obscure features hypothesis for innovation: One key to improving human innovation. (Unpublished doctoral dissertation.) University of Massachusetts, Amherst, MA.
McCaffery, T. (2012). Innovation relies on the obscure: A key to overcoming the classic problem of functional fixedness. *Psychological Science*, 12, 215–158.
McMullen, J.S., and Shepherd, D.A. (2006). Entrepreneurial action and the role of uncertainty in meaningful patterns: Evidence from comparisons of novice and experienced entrepreneurs. *Management Science*, 52, 1331–1344.
Miner, J.B., and Raju, N.S. (2004). When science divests itself of its conservative stance: the case of risk propensity differences between entrepreneurs and managers. *Journal of Applied Psychology*, 89, 3–13.
Mueller, J.S., Melwani, S., and Goncalo, J.A. (2012). The bias against creativity: why people desire but reject creative ideas. *Psychological Science*, 23, 13–17.
O'Brien, E., and Ellsworth, P.C. (2012). Saving the last for best: A positivity bias for end experiences. *Psychological Science*, 23, 163–165.
Sharot, T. (2011). *The Optimism Bias: A tour of the irrationally positive brain*. New York: Random House.
Sternberg, R.J., and Sternberg, K. (2011). *Cognitive Psychology*. Cincinnati: Cengage.
Tang, J., Kacmar, K.M., and Busenitz, L. (2012). Entrepreneurial alertness in the pursuit of new opportunities. *Journal of Business Venturing*, 27, 77–94.
Topolinski, S., and Struck, F. (2009). Incubation benefits only after people have been misdirected. *Consciousness and Cognition*, 18, 608–618.
Ward, T.B. (2004). Cognition, creativity, and entrepreneurship. *Journal of Business Venturing*, 19, 173–188.

4 From desire to achievement: the crucial role of self-regulation

Chapter outline

Self-regulation: its basic nature
Why self-regulatory skills may be especially valuable to entrepreneurs
Self-control: doing what we should do, and refraining from doing what we should not
Focus and persistence: having clear goals—and working consistently to reach them
Managing emotions and restraining impulses: why waiting is often better
Regulating impulsivity: resisting the urge to do it *now*
Delay of gratification: trading time for greater rewards
Metacognition: understanding and directing our own thoughts

* * *

> You have to learn the rules of the game. And then you have to play better than anyone else.
> (Albert Einstein)

> I never said it would be easy, I only said it would be worth it.
> (Mae West)

> Desire is the key to motivation, but it's the determination and commitment to an unrelenting pursuit of your goal – a commitment to excellence – that will enable you to attain the success you seek.
> (Mario Andretti)

Although the quotations above reflect the ideas of very diverse people, they are all, in a sense, concerned with *success*. In general terms, this is something nearly everyone wants, but what, precisely *is* it? Even a moment's reflection suggests that there may not be a single meaning to this term. What we seek may differ greatly across different activities or aspects of our lives. In *business*—and to a great extent, in *entrepreneurship*—success is often measured in terms of wealth and fame; in the *professions* (e.g., medicine, law), it may lie in a high income plus the esteem of one's colleagues; in the *arts*, critical acclaim may be crucial, while in *science* success is often measured in terms of publications in prestigious journals or winning important prizes. In our *personal lives*, in contrast, success centers on attaining the love, happiness, and meaningfulness we crave. Whatever the specific goals, though, almost everyone has—somewhere in the depths of her or his soul—visions of an alluring future in which their innermost desires and ambitions have been met.

Together, these points raise another intriguing question: Is there a best—or at least, *better* way—to attain success—to move from dreams or desires to their actual fulfillment? In a sense, that is what this book is all about. The key goal is to uncover, and describe, the tools (skills, experience, knowledge, characteristics, etc.) that entrepreneurs need to achieve *their* goals, which, as has been noted repeatedly, center on converting their ideas or visions into reality. In this chapter, we will focus on what, growing evidence suggests, is a very crucial ingredient in this equation: *self-regulation*.

Self-regulation: its basic nature

In essence, the term *self-regulation* refers to a collection of skills and capabilities individuals use to select key goals, monitor progress toward them, adjust their behavior (their thoughts, emotions, and actions) so as to enhance such progress, and—in a basic sense—*take their lives and fates actively into their own hands*. What are these skills and capabilities? Research findings, reviewed in detail in later discussions, reveal that among the most important are these:

1. Exerting self-control: performing actions that facilitate progress toward important goals (even if they are not enjoyable) while refraining from actions that are enjoyable but impede progress.
2. Demonstrating a combination of focus and persistence (staying focused on key goals and working persistently toward them).
3. The capacity to manage our emotions and impulses—for instance, to avoid hasty or rash behavior and decisions, and to delay gratification to times when it can be maximized.
4. Developing accurate *metacognition*—acquiring insightful self-knowledge, which involves monitoring and regulating our own cognitive processes so that they help us attain progress toward key goals, and also developing, and using, specific capabilities such as "knowing what we know and do not know".

Why self-regulatory skills are especially valuable for entrepreneurs

Clearly, these skills are important for everyone, but there are several reasons why they are especially valuable for entrepreneurs. First, in their efforts to create something new, entrepreneurs generally face situations in which the external rules or norms that guide behavior in many contexts are generally lacking. Unless they actively seek guidance from an experienced mentor, they have no direct supervisors, teachers, or other persons who evaluate their performance, provide feedback, and advise them on how to improve. Further, since they are, by definition, attempting to create something new, they often face situations in which they must, in a sense, "Make it up as they go along." Yes, venture capitalists often provide guidance, either directly or by encouraging contacts with knowledgeable people, but in many instances, such assistance is lacking. As a result, entrepreneurs often must rely primarily on their own skills and knowledge to help choose the paths that will help them advance toward their major goals.

Second, and especially for first-time entrepreneurs, they often find themselves performing tasks and filling roles they have not previously performed or occupied. As a result, it is especially crucial for entrepreneurs to accurately recognize what they know and do not know, so that they will seek help when it is truly needed.

Finally, as we will see in later chapters, entrepreneurs, as a group, are very high in optimism and in positive affect (i.e., positive moods and emotions). Although these tendencies are often beneficial, they can also generate harmful effects (e.g., a tendency to ignore relevant negative information; an inclination to engage in quick-and-dirty heuristic thinking when more systematic, careful analysis is essential; Baron et al., 2012) and a tendency to act *now* when, in fact, delay and reflection would be more effective. Again, because entrepreneurs usually operate in situations where external factors that might restrain these tendencies are absent (e.g., they lack supervisors who might tend to offset excessive optimism or excessive enthusiasm), they must rely on their own self-regulatory processes to hold these potentially damaging tendencies in check.

In short, entrepreneurs are, in several key respects, very much "on their own", and for this reason, must depend, more than others (e.g., people employed in large organizations), on their own self-regulatory mechanisms for guidance. For this reason, such skills take on special importance for them in terms of charting an effective course through largely unknown, and frequently turbulent waters (see Figure 4.1).

Having made what we hope is a convincing case for the special importance of self-regulatory skills, we will now examine evidence pointing to the powerful effects, and potential benefits, of these skills.

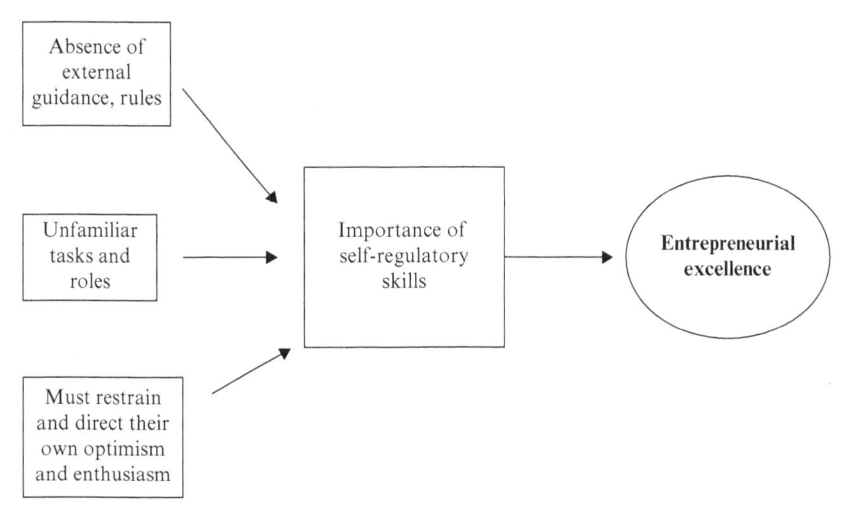

Note: Because they are attempting to create something new and better, entrepreneurs (especially first-time entrepreneurs) often face situations in which they have little external guidance and few, if any, established rules to follow. In addition, they are often performing tasks and filling roles they have not previously performed or occupied. They must also be capable of restraining their own high levels of optimism and enthusiasm, because if unchecked, these tendencies can lead to many difficulties.

Figure 4.1 *Why self-regulation may be especially important for entrepreneurs*

So strong and consistent is this evidence, that experts on self-regulation (Forgas et al., 2009, p. 5) have gone so far as to note that decades of research on the factors that influence success seem to converge on the conclusion that only two are important across every activity in which individual seek success: *intelligence* and *self-regulation*. After more than a century of careful study, we know quite a lot about intelligence, and one thing we know is that it is partly shaped by genetic factors and early childhood factors, both of which are not directly under our control. While that in no way implies that it cannot be changed, the fact that intelligence has a strong genetic component makes the task of doing so much more challenging. In contrast, self-regulation involves a number of different skills which can be readily acquired and strengthened—if individuals are willing to exert the effort to do so. This suggests that self-regulation offers a more practical route to achieving the personal success and fulfillment we seek. In support of this view, existing evidence (and there is a lot of it!) indicates that people who develop and practice effective self-regulation are often the ones who succeed—the ones who actually manage to convert their dreams into reality. Such persons share strong motivation for success and accomplishment with many others, but in contrast to most persons, they also possess the tools needed to convert this motivation and energy into actual progress. They exercise the kind of self-discipline, discretion, and direction that together help them move toward the goals they seek, whatever these may be (Baumeister and Tierney, 2011).

Self-control: doing what we should do, and refraining from doing what we should not

There is no doubt that we all find certain actions or activities much more enjoyable than others; and sadly, it is often the ones that are bad for us—in the sense of interfering with achieving our key goals—that are the most attractive. Unfortunately, as Mark Twain put it: "There is a charm about the forbidden that makes it unspeakably desirable." A key task we all face, then, is somehow regulating our behavior so that we do the things we should do (exercise regularly, eat moderately, save for a rainy day) and avoid the things we know, full well, we should not do (e.g., become couch potatoes, eat high calorie snacks or desserts, spend on impulse, etc.). The everyday term for this capacity is willpower, but another—the one generally used in research on this aspect of self-regulation—is *self-control* (de Ridder et al., 2012). Research on self-control has continued for several decades, but has recently become much more sophisticated—and informative. One clear finding of such research is that people differ greatly in the capacity to exert self-control (Tangney et al., 2004). Some are able to resist even powerful temptations to engage in actions, have thoughts, or experience emotions that interfere with their plans and goals, while others are much lower in this capacity (Baumeister and Alquist, 2009).

Clearly, self-control is an important aspect of self-regulation and plays a key role in success in almost any field or career. In fact, the capacity to exert self-control has been found to be strongly related to achieving real excellence in activities ranging from sports, and music through medicine, chess, art, and science. Generally, research on expertise indicates that to rise above the ordinary and become one of the top-performing people in any field, individuals must invest a minimum of 10,000 hours in highly focused and highly effortful deliberate practice (e.g., Campitelli and Gobet, 2011; Ericsson, 2006). Additional research has provided illuminating insights into the nature of this skill (it is indeed a skill because self-control can be strengthened through procedures described below).

One important finding with respect to self-control is that it appears to be an exhaustible cognitive resource; after individuals exert it in one situation or with respect to one activity (or temptation), their capacity to exert it again declines. Interestingly, experiencing positive feelings can restore self-control that has been depleted. For instance, Tice et al. (2007) conducted research in which individuals were placed in a situation where they had to resist strong temptation. Then, they were exposed to a second temptation. In between, one group of participants watched videos that had little or no impact on their moods, while others watched amusing videos that induced positive affect. Those in the second group were better at resisting a second temptation than those in the first (control) group. In addition, consuming glucose—a form of sugar that is quickly absorbed and used by the brain—has been found to exert similar effects (Baumeister and Alquist, 2009)—it too, can boost self-control. It may also be the case, though, that exerting self-control involves not so much a draining of cognitive capacity, as a shifting of attention between different goals (Fujita, 2011)—that issue is still being debated.

Is self-control concerned primarily with doing what we should do (but perhaps do not want to do) or refraining from doing what we should not do (but perhaps want to do)? A review of existing evidence indicates that it operates in both ways (de Ridder et al., 2012). So it both helps individuals to perform actions that assist them in reaching their important goals, and assists them in refraining from actions that block or interfere with progress toward these goals Further, it appears that people who are high in dispositional self-control (the tendency to exert self-control in many different situations) are especially effective at acquiring "good habits". For such persons, like everyone else, engaging in actions that are not enjoyable but facilitate goal achievement initially requires the exercise of self-control. However, very quickly, these actions become "automatic", so that they can be performed without further draining cognitive resources.

How important is self-control? In a sense, it is truly central to our efforts to regulate our own behavior, emotions, and cognition (Baumeister and Tierney, 2011). People high in self-control enjoy better physical and psychological health, better social relationships, greater career success, and higher levels of personal happiness than people low in self-control. Further, people low in self-control are more likely to engage in dangerous or deviant actions (using illegal drugs, abusing alcohol, aggressing against others, engaging in unprotected sex with strangers; Denson et al., 2012). Given these pervasive effects, it is not surprising to learn that self-control is an important factor in entrepreneurial excellence. It has recently been found to be positively related to both the performance of new ventures and entrepreneurs' personal subjective well-being (e.g., Baron et al., under review, 2012; Nambisan and Baron, in press, 2012). The basis for these relationships is complex, but in essence, reflects the fact that entrepreneurs, in their efforts to create something new and better, generally have no one—no supervisor or boss—to direct their behavior and keep them focused on tasks essential for moving toward their key goals. Instead, they must rely largely on their own self-control. Unfortunately, the temptations to avoid doing what is necessary but unappealing, and to do what is intrinsically enjoyable literally surround entrepreneurs. For instance, although most are very willing to work hard to develop and promote their new products or services, few enjoy the "managerial" side of entrepreneurship (e.g., keeping detailed financial records, hiring employees, assuring that taxes are paid and that government-established safety requirements are met). On the other hand, they may greatly enjoy the creative aspects of the process—and in fact may have a true passion for such activities (e.g., Cardon et al., 2009). For these and related reasons, a high level of self-control should be included in any list of essential tools for entrepreneurs.

Focus and persistence: having clear goals—and working consistently to reach them

Another important self-regulatory process involves two central components: (1) maintaining consistent interest in or focus on clearly defined long-term goals, and (2) demonstrating persistence in efforts to reach them (Duckworth et al., 2007). Together, these two components of self-regulation have been described by the term *Grit*, which is a synonym for tenacity and perseverance (Duckworth et al., 2007). Although these two components are somewhat distinct, they have been combined in recent research (e.g., Duckworth and Quinn, 2009; Duckworth et al., 2011), so we treat them here as a single aspect of self-regulation.

Both of these tendencies are related to self-control, but are also distinct from it in that self-control refers primarily to current actions (e.g., resisting an immediate impulse, behaving in ways inconsistent with long-term goals, or performing actions consistent with reaching such goals), while *Grit* focuses, instead, on processes that continue over extended periods of time. For instance, they are reflected in the long-term efforts to achieve the outstanding performance shown by experts in many different fields, and may provide a basic foundation of such performance (medicine, science, sports, music, etc.; Baron and Henry, 2010; see Figure 4.2).

Research evidence indicates that persons high in *Grit* (consistency of focus, persistence of effort) are significantly more successful than those relatively low in many different activities and careers. For instance, they achieve higher levels of education, are more successful in various professions, and more likely to excel in especially rigorous and difficult training programs (e.g., as cadets at West Point; Duckworth et al., 2007; Duckworth and Quinn, 2009; Duckworth et al., 2011). While the capacities to focus consistently and persistently on efforts to attain long-term goals are clearly relevant to achievement in many different contexts, they may be especially relevant to activities performed by entrepreneurs. Entrepreneurial activities rarely result in immediate rewards; in fact, the "payoffs" often emerge—if they ever do!—only months or years after the launch of a new

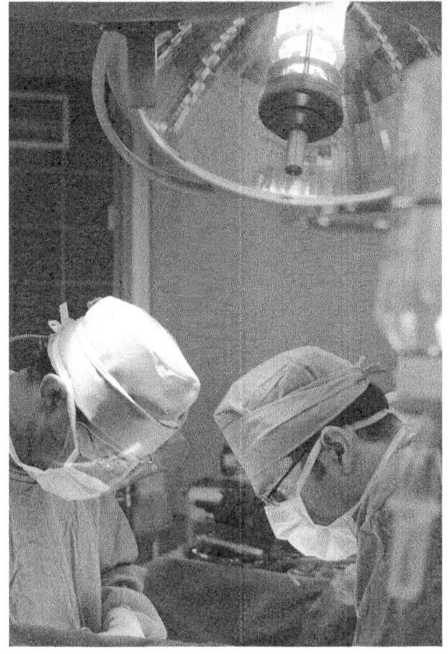

Note: A large body of evidence indicates that in order to acquire excellence in almost any field, individuals must focus on this goal for long periods of time (often, many years) and persistently exert effort to achieve it. In fact, in many fields, a "10,000 hour" rule seems to apply: a minimum of 10,000 hours of focused, effortful practice is necessary to achieve outstanding performance.

Sources: Fotolia 41513322; 2912982.

Figure 4.2 *The important role of consistent focus and persistent effort in attaining true excellence*

venture or other entrepreneurial activities. Yes, some entrepreneurs do reap large rewards quickly; for instance, the founders of Instagram, a photo-sharing company sold their new venture to Facebook for $1 billion less than two years after it was launched. But this is the exception rather than the rule, and most entrepreneurs must plug along for long periods of time before receiving any tangible rewards, or even drawing a salary. Further, they must remain focused on the core goals they are pursuing because their resources are generally too limited to permit them the luxury of shifting their direction or focus radically.

Recent research findings indicate that, in fact, *Grit* is significantly related to entrepreneurs' success. For instance, Nambisan and Baron (in press, 2012) found that among a group of high-tech entrepreneurs, *Grit* was positively related to new venture performance. Further, this relationship was influenced by self-control, being stronger when entrepreneurs were high in self-control than when they were low. This suggests that these two distinct aspects of self-regulation can combine to equip entrepreneurs with skills that greatly help them to make the possible real.

In sum, where entrepreneurial excellence is concerned, the prize goes not necessarily to the most imaginative or inspired individuals, but rather to the ones who select appropriate and attainable goals, remain focused on them, and work persistently over time for their achievement—and continue to do so even when this involves engaging in actions or activities they find unappealing.

Managing emotions and restraining impulses: why waiting is often better

Almost certainly, you have heard the term *emotional intelligence* at some time or other. This phrase was first popularized by Goleman (1995), who used it to refer to a cluster of capacities or skills related to the "feeling side of life". In Goleman's original proposals, emotional intelligence included (1) the ability to accurately perceive emotions in oneself and others, (2) to use emotions to facilitate cognitive activities such as thinking and problem solving, (3) to understand emotions and relationships among them, and perhaps most crucial (4) the capacity to manage emotions in both ourselves and others (Grewal and Salovey, 2005).

Initially, there was little evidence for the validity of these suggestions—for the view that people differ greatly in these skills, and that such differences influence their careers, success, and happiness. However, in recent years, a growing body of carefully conducted research has been performed to investigate the nature and impact of emotional intelligence, and much of this evidence offers support for at least a modified version of Goleman's ideas (e.g., Grewal and Salovey, 2005). Most important, perhaps, is the finding that several aspects of emotional intelligence are related to important life outcomes. For instance, people high in emotional intelligence have fewer negative but more positive interactions with friends and family and, in general, get along better with others in a wide range of contexts—work, school—virtually anywhere that people meet and interact (e.g., Lopes et al., 2003). Persons high in emotional intelligence also tend to be more successful in their careers or jobs than persons lower in these skills: they receive more and faster promotions, receive higher salaries, earn more from their own businesses, and express greater satisfaction with their work, whatever it is (e.g., Baron and Markman, 2003; Lopes et al., 2011).

Clearly, emotional intelligence (often abbreviated as EI) offers important benefits to the persons who are high in such skills, but once again, there are several reasons why EI—and especially the capacity to manage our own emotions and influence those of others—may be especially useful for entrepreneurs. First, as noted previously, entrepreneurs tend to be exceptionally high in positive affect; although this is often an asset, it can, at very high levels, become a liability, so it is crucial that entrepreneurs be aware of their own high levels of positive affect and develop the capacity to restrain them (Baron et al., 2012). Second, unless entrepreneurs can both accurately assess others' emotions and influence them, they may be unable to generate the enthusiasm needed to secure support and commitment for their new ventures or other entrepreneurial activities. Third, in order to persevere, entrepreneurs must be able to cope effectively with the intense negative feelings generated by business failure (see Chapter 8). Finally, accurate understanding of others' emotions and feelings is often crucial to developing effective relationships with them, and thus for establishing large and high-quality social networks—a factor that has been clearly identified as a key ingredient in entrepreneurial success (e.g., Barringer and Ireland, 2011; see Chapter 5).

Fortunately, the skills that combine to generate emotional intelligence can be developed—individuals can enhance their abilities to accurately recognize emotions (their own and others'), to put their emotions to good use (e.g., use them to maximize performance of various tasks), and to manage their emotions effectively, so that they facilitate rather than impede progress toward important goals (e.g., Grewal et al., 2006). One technique for improving these skills involves focusing attention on information that will induce the emotions individuals wish to experience, or block emotions they do not wish to experience. For instance, in order to reduce negative emotions after a setback, we can focus on positive aspects of the situation—what we have learned and how we can do better in the future. Similarly, in order to prepare ourselves for an unpleasant confrontation with another person, we can focus on information that causes us to experience anger—an emotion that can strengthen our resolve to stand firm. Recent research has reported that individuals who possess a good understanding of steps they can take to increase their own anger in such contexts do in fact demonstrate greater dominance in them (e.g., Cote et al., 2011).

Two other techniques for regulating emotions—especially negative ones—involve reappraising an unpleasant situation so as to minimize its negative nature, and distraction—thinking about other, less unpleasant situations or engaging in actions we enjoy. An example of reappraisal is provided by an entrepreneur who fails to secure a large order for her company, but then chooses not to dwell on her feelings of disappointment, but rather on the fact that she has now established a relationship with an important customer—one that may well yield orders in the future, if not now. An example of distraction would involve driving thoughts of the lost order from your mind by thinking, instead, of more positive events or outcomes (Sheppes et al., 2011), or simply engaging in enjoyable activities—to the extent that these are not viewed as being "bad" for us! All of these techniques can be readily learned and used, and to the extent entrepreneurs' practice using them, they may acquire yet another valuable self-regulatory mechanism.

Regulating impulsivity: resisting the urge to do it now

Warnings against behaving impulsively—quickly and without considering the possible consequences—are common, and with good reasons. Many impulsive actions are at least potentially harmful to the persons who perform them: impulse purchases, gambling, excessive drinking, unprotected, high-risk sex, speeding, and so on. Yet, people engage in these behaviors, and often do so with the full realization that they are dangerous or potentially harmful (e.g., Cyders et al., 2007). Such actions can be seen as ones in which current impulses are so strong that they overwhelm the self-regulatory mechanisms that would, in many cases, tend to restrain them (e.g., self-control). More specifically, they can be viewed as deriving from failures of emotional self-regulation. When individuals yield to such impulses they do often "go off the track" in terms of reaching important goals: for instance, they eat high-calorie foods while on a diet; they buy items they really cannot afford; and they embark on risky sexual adventures that promise both excitement and pleasure.

Unfortunately, additional research findings indicate that positive emotions often increase the likelihood that people will act impulsively (e.g., Tice, 2009). Such feelings tend to reduce attention to negative information—including possible harm that may result from impulsive actions—and shift priorities toward attaining immediate rewards, even if these carry considerable risk. As noted previously, entrepreneurs tend to be very high in dispositional positive affect—the tendency to experience positive moods or feelings often and in many different situations, (e.g., Baron et al., 2012). In addition, they also tend to be high in a characteristic known as *locomotion*—the desire to move forward now rather than delay (e.g., Kruglanski et al., 2007). Together, these tendencies may make them especially susceptible to the temptation of strong impulses. That is another reason why regulation of emotions is indeed a key skill for entrepreneurs, one they should seek to develop, and one that provides an important tool for attaining entrepreneurial excellence.

Delay of gratification: trading time for greater rewards

In the early 1960s Walter Mischel, at the time a young psychologist, performed a series of experiments with three- and four-year-old children. He placed a marshmallow in front of each child and told them "You can have this one right now, but if you wait until I come back, you can have two. If you want me to come back right away, though, you can ring this bell." Then, he left the room and waited to see what the children would do. As you can probably guess, some yielded to temptation at once, and ate the marshmallow—some, even before Mischel could leave the room. Others, however, delayed. Some managed to wait only a few seconds, and then rang the bell. But others waited until Mischel returned, 15 minutes later, because they wanted to receive the larger reward—two marshmallows instead of one. This simple experiment, which has been repeated many times with various methods, and with adults as well as children—demonstrates clearly that people differ greatly in their capacity to delay gratification—to refrain from enjoying small rewards now in order to attain larger ones at a later time (Mischel, 1974, 1977). In a sense, recognition of this skill lies behind the advice offered by Dave Ramsey—the famous American financial advisor, who often berates people who call his radio show for living beyond (instead of beneath) their means, and for not saving for the future. Basically, he criticizes them for being unable to delay gratification—being unable to wait until the right time to begin enjoying various rewards, such as a fancy car or a larger house.

Is the capacity to delay receiving rewards really an important aspect of self-regulation? When Mischel studied the lives of the children in his early research, his findings were dramatic: those who were good at delaying gratification—resisting the siren call of one marshmallow in order to obtain two later—differed greatly from those who showed lower capacity to delay. The children who were already good at delaying gratification at age four experienced greater success in school, in their careers (decades later), and formed more stable personal relationships than those low in this skill. The latter group were also more likely to experience problems in school and in their personal lives, often had difficulty paying attention while performing various tasks, and had difficulties forming and maintaining friendships. Perhaps most dramatically, the high-delayers had S.A.T. scores 210 points higher than those low in the capacity to delay (Mischel et al., 2011).

Note: Individuals on a diet should avoid focusing on foods such as these. Instead, they should try to distract their own attention away from them. That is one basic technique for delaying gratification.

Source: Fotolia 27333208.

Figure 4.3 *Delay of gratification: how not to do it!*

Further research provides insights into how people manage to refrain from enjoying immediately available rewards. In particular, they are good at regulating their own attention so that, for instance, they shift their thoughts away from the rewards available now to other topics (see Figure 4.3). The four-year-olds who managed to wait for Professor Mischel's return, for instance, would cover their eyes, turn away from the tray containing the marshmallow, play with their hair—almost anything to help shift their own attention away from the treat in front of them. Research with adults indicates that similar—but more sophisticated—processes are at work. Adults, too, shift their attention to other topics and use their much more highly developed mental processes to accomplish this task. For instance, they may focus on the future, and imagine the rewards they will attain then; and of course, if they have their cell phones available, they can quickly divert their thoughts to a wide range of other topics. If possible, adults attempting to delay gratification will also remove themselves physically from temptations for immediate indulgence—for cakes in their windows or to avoid looking at them! (See Figure 4.3.) When Odysseus, the famous hero of ancient Greece, had himself tied to the mast of his ship it was so that he could hear the song of the sirens (mythical creatures who lured sailors to their death with irresistible calls), but could not obey them.

While the capacity to delay gratification is important in many contexts and situations, it again seems especially crucial for entrepreneurs. For instance, when a new venture begins to yield cash flow, there may be strong temptation to distribute these gains to founders, investors, and others, since these persons have already waited a long time for such benefits. But almost without exception, it is better to resist such temptations and to reinvest these initial returns in the business—in the hope, of course, of far larger returns in the future. In fact, entrepreneurs must often exist in a state of near-impoverishment for months or even years, and yielding to temptations to improve their own or investors' standards of living is akin to a farmer deciding to eat the seeds for next year's crops. Fortunately, once again, individuals can enhance their own capacity to delay gratification by several techniques—most based on shifting attention away from the current temptations. As Mischel (2009) puts it: "Once you realize that will power is just a matter of learning how to control your attention and thoughts, you can really begin to increase it" (cited in Lehrer, 2009).

Metacognition: understanding and regulating our own thoughts

When the children in Mischel's experiments covered their eyes and looked away from the treats on the table in front of them, they were trying to regulate their own cognition. Similarly, when adults avoid having snacks they love in the house, or when they count to ten when angered by another person, they, too, are attempting to regulate their own thoughts (and perhaps their emotions, too). These incidents illustrate another important aspect of self-regulation known as *metacognition*, which can be defined as thoughts, beliefs, and other cognitive processes used to assess and control our own cognitions—our own thinking (e.g., Hertzog and Dunlosky, 2011). More generally, it refers to awareness of and efforts to regulate our own cognitive processes (Flavell, 1979).

Although metacognitive abilities develop early in life—they have been observed among children as young as 18 months old (Lehrer, 2009)—they are not a fixed or unchangeable characteristic: rather, as is true of other aspects of self-regulation, metacognition can be viewed as a skill or set of skills that can be acquired or strengthened (e.g., Schmidt and Ford, 2003). Interestingly, although some aspects of our cognitive systems do decline to a degree with age (e.g., entering information into certain kinds of memory slows down and become more difficult), metacognitive capacity seems to remain relatively constant. That means that even if we become less effective at entering information into memory or retrieving information from it as we age, we can use metacognitive skills—insights into our own cognitive processes—to compensate for such declines (Hertzog and Dunlosky, 2011). That may be one reason why older and experienced individuals are highly valued in many cultures, even though the rate at which they can acquire new information decreases, they develop compensatory mechanisms for drawing on the vast store of knowledge and wisdom they have already acquired.

Metacognition includes several different components, but here, we will focus on two that are especially relevant to entrepreneurship: metacognitive awareness and metacognitive resources (knowledge and experience). Metacognitive awareness refers to awareness of one's own cognitions—what individuals know about themselves as cognitive processors of information. For instance, an individual high in metacognitive awareness might recognize that he or she is not very good at certain kinds of tasks (e.g., keeping track of many small details), but quite good at others (e.g., recognizing major themes in complex sets of data). Metacognitive resources, in contrast, refer to the knowledge and experience that individuals draw upon in devising or selecting the most effective cognitive strategies to employ in a given situation (Flavell, 1987). For instance, suppose that an entrepreneur experiences unexpected difficulties in terms of product development. Should efforts to resolve these problems proceed through trial-and-error (the strategy Edison used in perfecting the electric light bulb), by applying principles gained in other contexts (e.g., information about the ability of various substances to survive exposure to high temperatures), or through some other tactic? Both aspects of metacognition can assist individuals in choosing or developing the most appropriate strategies to employ in performing important tasks involving cognition, for example, solving complex problems or adjusting effectively to rapidly changing environmental conditions—ones often faced by entrepreneurs.

Recently proposed theoretical frameworks (e.g., Haynie et al., 2010; Haynie and Shepherd, 2009) suggest that both entrepreneurs' metacognitive awareness and metacognitive resources are important in adopting cognitive strategies that lead to desirable outcomes relating to specific entrepreneurial goals. Further, evidence reported recently by Baron et al. (under review, 2012) indicates that one aspect of metacognitive knowledge—knowing when to withdraw from a failing course of action—has significant effects on the strategies founding entrepreneurs choose for their new ventures.

One aspect of metacognition that is especially important for entrepreneurs might be described as "knowing what we know and do not know". This implies being aware of our stores of knowledge, our understanding of the limits of such information, and our ability to determine how best to

Table 4.1 *Items for measuring an important aspect of metacognition*

Q: To What extent is each of the following statements like you?

	Not at all like me	*Not like me*	*Neutral*	*Like me*	*Just like me*
I have no difficulty admitting that I do not know something, or do not know how to perform some task.	❍	❍	❍	❍	❍
I am confident that I know what I need to know to run my business (practice, job) effectively.	❍	❍	❍	❍	❍
I find it easy to ask for information, help, or advice from others.	❍	❍	❍	❍	❍
I can usually tell if I have sufficient knowledge/skills to perform tasks I take on.	❍	❍	❍	❍	❍
I am generally aware of what I know or don't know in any situation.	❍	❍	❍	❍	❍
If I don't have enough knowledge or experience to tackle a task, I don't tackle it.	❍	❍	❍	❍	❍

Note: The items shown here are designed to measure a key aspect of metacognition—knowing what we know and what we do not know.

Source: Courtesy of R.A. Baron.

use it. Recent evidence collected by Nambisan and Baron (in press, 2012) indicates that this is a crucial skill for entrepreneurs. They found that a measure of this aspect of metacognition (see Table 4.1) was significantly related both to the performance of new ventures and to entrepreneurs' subjective well-being—how happy they were with important aspects of their personal lives. These findings suggest that clear understanding of what we know and do not know is a key ingredient in developing entrepreneurial excellence. Here is an example from my own life.

Many years ago, I was a visiting faculty member at the University of Texas. During my time in Austin, I discovered a new style of barbecue which I found to be delicious: it involved smoking various cuts of meat over fires for many hours, fires fueled, in part, by mesquite, a plant native to large areas of Texas. The final result was tender, smoky, and had a unique flavor. These culinary experiences led me to wonder if a restaurant serving real Texas-style barbecue could be successful in other locations and whether, for instance, I could open such a business in W. Lafayette,

Indiana, when I returned to my regular position at Purdue University. I concluded that such a restaurant could be a tremendous success—a veritable gold mine, given the 30,000+ often hungry (!) students in town. But the more I thought about it, the more I realized that I did not really have the knowledge needed to open and run such a restaurant. What kind of equipment would I need? How would I obtain mesquite, to produce the unique flavor I loved so much? How could I staff the restaurant? I knew that I could probably acquire all this information and the necessary skills, but that this would mean interrupting or even ending my academic career. In short, I realized that I did not know what I needed to know and that, consequently, this was not a potential opportunity I should pursue. My metacognitive skills—although certainly flawed in many ways—were effective enough to help me reach what was, for me, the correct decision at the time. Indeed, it is now widely recognized in the field of entrepreneurship that recognizing an opportunity is just the first step in a process that must also involve entrepreneurs asking themselves "Is this an opportunity I can actually develop?" (McMullen and Shepherd, 2006). Clearly, then, metacognition—both metacognitive awareness (which includes understanding the limits of our own knowledge and skills) and metacognitive resources (knowing how to apply these to regulating our own cognition) are important skills that, along with other aspects of self-regulation, can contribute strongly to entrepreneurs' excellence in imaging, and developing something new and better.

The key point of this chapter can be readily stated: several key aspects of self-regulation—the cognitive processes that help us to monitor and direct our own actions, thoughts, and emotions—are crucial for success in many activities. However, because of the unique conditions they confront, including the lack of external direction, rules, or guidelines, these inner cognitive resources may be especially relevant for entrepreneurs who must truly "make it up as they go along", and identify their own unique path to their ultimate, overarching goals.

Summary of key points

The term self-regulation refers to a collection of skills and capabilities individuals use to select key goals, monitor progress toward them, and adjust their behavior (their thoughts, emotions, and actions) so as to enhance such progress. These skills are important in many situations, but are especially crucial for entrepreneurs since they often lack external rules or guides for their behavior. They vary greatly from one person to another. One important aspect of self-regulation is *self-control*, which involves regulating our behavior so that we do the things we should do (i.e., ones that help us progress toward our key goals), but refrain from doing the things we should not do (ones that impair goal progress). Self-control has been found to be strongly related to success in many careers or activities, and has recently been found to be strongly related both to the performance of new ventures and entrepreneurs' subjective well-being. Self-control varies greatly across individuals, and appears to be a depletable resource, but can be restored by increased positive affect, among other things.

Another key aspect of self-regulation (known as *Grit*) involves the capacities to stay focused on key goals and to work persistently, over time, to attain them. These capacities, too, have been found to be strongly related to entrepreneurs' success. Regulating emotions and impulses is yet another aspect of self-regulation; individuals lacking in such skills often engage in hasty or rash actions that can interfere with their progress toward key goals. Finally, self-regulation also involves *metacognition*—thoughts, beliefs, and other cognitive processes used to assess and control our own cognition, our own thinking. An especially crucial aspect of metacognition for entrepreneurs is the capacity to recognize "what they know and do not know". This capacity seems to be essential from the point of view of helping entrepreneurs to choose opportunities that they are, in fact, competent to develop, while avoiding ones that are outside their knowledge or capabilities.

References

Baron, R.A., and Henry, R.A. (2010). How entrepreneurs acquire the capacity to excel: Insights from basic research on expert performance. *Strategic Entrepreneurship Journal*, 4, 49–65.

Baron, R.A., and Markman, G.D. (2003). Beyond social capital: The role of entrepreneurs' social competence in their financial success. *Journal of Business Venturing*, 18, 41–60.

Baron, R.A., and Nambisan, S. (under review). Self-regulatory processes and the two "sides" of entrepreneurial success: Firm-level performance and personal well-being.

Baron, R.A., Hmieleski, K.M., and Henry, R.A. (2012). Entrepreneurs' dispositional positive affect: The potential benefits—and potential costs—of being "up". *Journal of Business Venturing*, 27, 310–324.

Baron, R.A., Hmieleski, K.M., Fox, C., and Casper, C. (under review, 2012). Entrepreneurs' self-regulatory processes and the adoption of high-risk strategies: Effects of self-control and metacognitive knowledge.

Barringer, B., and Ireland, D. (2011). *Entrepreneurship*, 4th edn. Upper Saddle River, NJ: Pearson Education.

Baumeister, R.F., and Alquist, J.L. (2009). Self-regulation as a limited resource: Strength mode of control and depletion. In J.P. Forgas, R.F. Baumeister, and D.M. Tice (eds), *Psychology of Self-regulation: Cognitive, affective, and motivational processes*, pp. 21–35. New York: Taylor & Francis.

Baumeister, R.F., and Tierney, J. (2011). *Willpower*. New York: The Penguin Press.

Baumeister, R.F., Vohs, K.D., and Tice, D.M. (2007). The strength model of self-control. *Current Directions in Psychological Science*, 16, 351–355.

Campitelli, G., and Gobet, F. (2011). Deliberate practice: Necessary but not sufficient. *Current Directions in Psychological Science*, 20, 280–285.

Cardon, M.S., Wincent, J., Sing, J., and Drnvosek, M. (2009). The nature and experience of entrepreneurial passion. *Academy of Management Review*, 34, 511–532.

Cote, S., DeCelles, K.A., McCarthy, J.M., Van Kleef, G.A., and Hideg, I. (2011). The Jekyll and Hyde of emotional intelligence: Emotion-regulation knowledge facilitates both prosocial and interpersonally deviant behavior. *Psychological Science*, 11, 1073–1080.

Cyders, M.A., Smith, G.T., Fischer, S., Annus, A.M., and Petgerson, C. (2007). Integration of impulsivity and positive mood to predict risky behavior: Development and validation of a measure of positive urgency. *Psychological Assessment*, 29, 107–118.

Denson, T.F., DeWall, C.N., and Finkel, E.J. (2012). Self-control and aggression. *Psychological Science*, 21, 20–25.

De Ridder, D., Lansvelt-Mulders, G., Finkenauser, C., Stok, F.M., and Baumeriter, R.F. (2012). Taking stock of self-control: A meta-analysis of how trait self-control; relates to a wide range of behaviors. *Personality and Social Psychology Bulletin*, 16, 76–99.

Duckworth, A.L., and Kern, M.L. (2011). A meta-analysis of the convergent validity of self-control measures. *Journal of Research in Personality*, 45, 250–268.

Duckworth A.L., and Quinn, P.D. (2009). Development and validation of the short Grit Scale (Grit-S). *Journal of Personality Assessment*, 91, 166–174.

Duckworth, A.L., Peterson, C., Matthews, M.D., and Kelly, D.R. (2007). Grit: Perseverance and passion for long-term goals. *Journal of Personality and Social Psychology*, 92, 1087–1101.

Duckworth, A.L., Quinn, P.D., and Seligman, M.E.P. (2009). Positive predictors of teacher effectiveness. *Journal of Positive Psychology*, 19, 540–547.

Ericsson, K.A. (2006). The influence of experience and deliberate practice on the development of superior expert performance. In K.A. Ericsson, N. Charness, R. Hoffman, and J. Feltovich (eds), *The Cambridge*

Handbook of Expertise and Expert Performance, pp. 683–703. New York: Cambridge University Press.

Flavell, J. (1979). Metacognition and cognitive monitoring: a new area of cognitive-developmental inquiry. *American Psychologist*, 34, 906–911.

Flavell, J. (1987). Speculations about the nature and development of metacognition. In F.E. Weinert, and R.H. Kluwe (eds), *Metacognition, Motivation, and Understanding*. Hillside, NJ: Erlbaum.

Forgas, J.P., Baumeister, R.F., and Tice, D.M. (2009). *The Psychology of Self-Regulation: Cognitive, Affective, and Motivational Processes*. New York: Psychology Press.

Fujita, K. (2011). On conceptualizing self-control as more than the effortful inhibition of impulses. *Personality and Social Psychology Review*, 155, 342–366.

Goleman, D. (1995). *Emotional Intelligence: Why it can matter more than IQ*. New York: Bantam Books.

Grewal, D., and Salovey, P. (2005). Feeling smart: The science of emotional intelligence. *American Psychologist*, 93, 330–339.

Grewal, D.D., Brackett, M., and Salovey, P. (2006). Emotional intelligence and the self-regulation of affect. In D.K. Snyder, J.A. Simpson, and J.N. Hughes (eds), *Emotion Regulation in Couples and Families*, pp. 37–55. Washington, DC: American Psychological Association.

Haynie, J.M., and Shepherd, D.A. (2009). A measure of adaptive cognition for entrepreneurship research. *Entrepreneurship Theory and Practice*, 33, 695–734.

Haynie, J.M., Shepherd, D., Mosakowski, E., and Early, P.C. (2010). A situated metacognitive model of the entrepreneurial mindset. *Journal of Business Venturing*, 25, 217–229.

Hertzog, C., and Dunlosky, J. (2011). Metacognition in later adulthood: spared monitoring can benefit older adults' self-regulation. *Current Directions in Psychological Science*, 10, 167–173.

Kruglanski, A.W., Pierro, A., Higgins, E.T., and Capozza, D.L. (2007). "On the move" or "staying put": Locomotion, need for closure, and reactions to organizational change. *Journal of Applied Social Psychology*, 37, 1305–1340.

Lehrer, J. (2009). Don't! The secret of self-control. *The New Yorker*, 18 May.

Lopes, P.N., Salovey, P., and Straus, R. (2003). Emotional intelligence, personality, and the perceived quality of social relationships. *Personality and Individual Differences*, 35, 641–658.

Lopes, P.N., Nezlek, J.B., Extremera, N., Hertel, J., Fernández-Berrocal, P., Schütz, A., and Salovey, P. (2011). Emotion regulation and the quality of social interaction: Does the ability to evaluate emotional situations and identify effective responses matter? *Journal of Personality*, 79, 429–467.

McMullen, J.S. and Shepherd, D.A. (2006). Entrepreneurial action and the role of uncertainty in the theory of the entrepreneur. *Academy of Management Review*, 31, 132–152.

Mischel, W. (1974). Processing delay of gratification. In Berkowitz, L. (ed.), *Advances in Experimental Social Psychology*, Vol. 7, pp. 249–292. New York: Academic Press.

Mischel, W. (1977). The interaction of person and situation. In D. Magnusson, and N.S. Endler (eds), *Personality at the Crossroads: Current Issues in Interactional Psychology*, pp. 333–352. Hillsdale, NJ: Lawrence Erlbaum Associates.

Mischel, W., Ayduk, O., Berman, M., Casey, B.J., Gotlib, I., Jonides, J., Kross, E., Wilson, N., Zayas, V., and Shoda, Y. (2011). "Willpower" over the life span: Decomposing self-regulation. *Social Cognitive and Affective Neuroscience*, 6, 252–256.

Nambisan, S., and Baron, R.A. (in press, 2012). Entrepreneurship in innovation ecosystems: entrepreneur's self-regulatory processes and their implications for new venture success. *Entrepreneurship Theory and Practice*.

Schmidt, A.M., and Ford, J.K. (2003). Learning within a learner control training environment: The interactive effects of goal orientation and metacognitive instruction on learning outcomes. *Personnel Psychology*, 56, 405–429.

Sheppes, G., Scheibe, S., Gaurav, S., and Gross, J.J. (2011). Emotion-regulation choice. *Psychological Science*, 22, 1391–1396.

Tangney, J.P., Baumeister, R.F., and Boone, A.L. (2004). High self-control predicts good adjustment, less pathology, better grades, and interpersonal success. *Journal of Personality*, 72, 271–324.

Tice, D.M. (2009). How emotions affect self-regulation. In Forgas, J.P., Baumeister, R.F., and Tice, D.M. (eds), *Psychology of Self-regulation*, pp. 201–215. New York: Psychology Press.

Tice, D.M., Baumeister, R.F., Shmueli, D., and Muraven, M. (2007). Restoring the self: Positive affect helps improve self-regulation following ego depletion. *Journal of Experimental Social Psychology*, 43, 379–384.

5 The social side of entrepreneurship: getting the help you need

Chapter outline

Getting off to a strong start: building a "dream" founding team
What makes for an effective founding team? Similarity or complementarity?
Building social networks—and using them successfully
The nature and impact of social networks—and why they are crucial for entrepreneurs
Social capital: the benefits derived from social networks
Beyond social capital: the importance of social/political effectiveness
The nature of political and social skills
Political skills and the development—and utilization—of social networks
Do social and political skills play an important role in entrepreneurship? Evidence indicating that they do

* * *

> Individuals don't win in business, teams do.
> (Sam Walton)

> If everyone is moving forward together, then success takes care of itself.
> (Henry Ford)

> Individual commitment to a group effort – that is what makes a team work, a company work, a society work, a civilization work.
> (Vince Lombardi)

There are some activities people do best alone—studying, writing, and practicing a new skill. Entrepreneurship, however, is *not* one of them. Most new ventures are founded by teams rather than individuals, and in general, efforts to introduce something new and better in almost any context usually involve the joint or coordinated efforts of several persons. Why? Because individual entrepreneurs rarely possess all the information, knowledge, or resources they need to convert the products of their own creativity into something tangible—a new company, new service, product, new way of doing things . . . or anything else new and better (e.g., Bledo et al., in press, 2012). In fact, they often

Note: In recent years, many universities have established incubator centers or special entrepreneurship centers, facilities designed to help faculty and students, as well as people in nearby communities, in their efforts to engage in entrepreneurial activities.

Source: Robert A. Baron.

Figure 5.1 *University entrepreneurship centers: one important means for assisting entrepreneurs*

need help from others—help ranging from advice and technical assistance, through concrete financial resources and emotional support. Recognition of this point lies behind the fact that many universities have established "Incubator Centers", or special Entrepreneurship Centers (see Figure 5.1) in which faculty and students receive a wide range of help in their efforts to act entrepreneurially—for instance, found a new venture—and in which special programs designed to assist current or potential entrepreneurs are often provided.

If entrepreneurs often cannot do it alone, two important questions arise: (1) What kinds of help do they need? And even more important, (2) How can they obtain it? Those are the issues on which we will focus in this chapter. The guiding principle behind all of the topics presented is that the more effective entrepreneurs are at attaining high-quality help—whether it involves information, social support, or financial resources—the more likely they are to attain the excellence—and success—they seek. Now, without further delay, we will turn to what the title of this chapter describes as "the social side of entrepreneurship"—recognition of the fact, so eloquently stated in the quotations above, that working effectively with others, and obtaining their support and assistance, is, indeed, one key ingredient in entrepreneurial excellence.

Getting off to a strong start: building a "dream" founding team

Earlier, it was noted that few entrepreneurs attempt the journey from imagination to reality alone. In fact, a high proportion of new ventures are started by *founding teams* consisting of several members, and within large organizations, efforts to encourage change and innovation too, often involve teams. True, in some settings—ones in which entrepreneurs do not attempt to start a new business or a new division or "spin-off" within an existing company—they *do* work as individuals. Overall, though, the task of making the *possible* real involves the joint efforts of several persons. This means that one of the most important initial steps entrepreneurs take is that of forming the founding team—the group of persons who will work together toward shared goals, whatever these happen to be—starting and running a profitable new company, doing social good, bringing a new product or service to the marketplace, or any other shared objectives. How do these teams form? And is there a best (or at least better) way to build them? Read on, because the answers to these seemingly straightforward questions may surprise—and greatly benefit—you!

What makes for an effective founding team? Similarity, or complementarity?

Think about the persons with whom you are close friends. What is the basis for your relationships with them? If you give this question some thought, you may realize that these friendships are, in many cases, based on similarity. You are similar to these people in terms of personal background—age, gender, race, religion, ethnicity, where you grew up, or in terms of interests values, and attitudes. You probably share various beliefs with them, work in similar occupations, and have similar education or training, and so on. This is hardly surprising: decades of research by psychologists and sociologists have found that similarity is one of the most powerful foundations of personal relationships. In fact, one prominent psychologist (Donn Byrne; e.g., Byrne and Griffitt, 1973) describes this as *the law of attraction*—the more similar to one another people are (in any of many different ways), the more they tend to like one another. In addition, they also tend to trust and feel comfortable around people similar to themselves in various ways (e.g., Rueff et al., 2003). Little wonder, then, that most founding teams of entrepreneurs are composed of persons who are highly similar to one another. In fact, over the years, I have met and worked with literally hundreds of such teams, and in general, and as research findings suggest, the members are highly similar to one another. For instance, among student teams, all the members may be engineers, or management students, or biologists; and they tend to be similar in age, family background, and dozens of other possible ways. Again, this is far from surprising: similarity facilitates communication—all team members speak "the same language", and it makes it easier to share basic goals. But is this a good basis for forming such teams? The answer provided by research findings is *definitely not*.

When people are highly similar to one another and share almost identical perspectives, they become subject to important forces that can interfere with the team's effectiveness. One of these is known as "Groupthink" (e.g., Turner and Pratkanis, 1998), and refers to the fact that teams that are highly cohesive—composed of highly similar persons who share mutual liking—may tend to focus more on agreeing with one another than on anything else. They come to believe that the views they share *must* be the correct ones, and that information from outside the group is neither to be desired nor trusted. As a result, they create a kind of "echo chamber", in which they hear only their own voices and views, and in which they become increasingly convinced that these views are correct. As the process has often been described, they come to believe that: "It must be so, since we all say it's so."

Why is this potentially dangerous to entrepreneurs? For one thing, it can lead founding teams to "fall in love" with their own ideas. Since they are relatively unwilling to listen to negative input from outside the group, they come to perceive their own ideas and plans for developing them as truly excellent—in need of no change or correction. As a result, they sometimes go off the "deep end" because, to put it simply, they "close ranks" and will not listen to good advice from people who are not members of the team.

Another potential danger is that highly homogeneous groups (ones in which all the members are similar in terms of background, attitudes, beliefs, and training) may be that such teams can be sadly lacking in various skills needed to start and run a new company—or to act entrepreneurially in other ways or contexts. For instance, consider a team that consists entirely of engineers—and not only engineers, but ones trained or training in the same branch of engineering (e.g., materials, mechanical, civil, computer engineering, whatever!). As a result, the team—although perhaps highly competent with respect to the engineering aspects of the idea they hope to develop, may know next to nothing about the business side of starting a new venture: nothing about finance, accounting, human resource issues, legal issues, marketing. The result? They may soon encounter major difficulties stemming from the fact that their high degree of similarity almost assures that although they have depth as a team in a few areas, they are lacking in breadth. The same would be true for any other founding team composed of members who share virtually identical backgrounds, interests, skills and knowledge.

A better approach to building founding teams, extensive research suggests, is to form them not on the basis of similarity (which is sometimes termed *homophily*), but on the basis of *complementarity*—the extent to which the individual members provide an array of non-overlapping skills, knowledge, and personal contacts which, together, add to the team's overall competence. In fact, research findings indicate that teams formed in this way are far superior in performance than ones based on similarity, proximity (members who happen to live or work in the same geographic location), or even social ties (i.e., the members know the same or overlapping people, even if they did not know each other previously; Rueff et al., 2003). Founding teams formed on the basis of complementarity are less likely to suffer from the kind of "groupthink" described above, and are more willing to seek outside advice and guidance—important ingredients in entrepreneurial success (e.g., Bae, 2012; Bonaccio and Dalal, 2006). Further, since they know different people (they have different personal contacts), this broadens the scope of team members' combined social networks. We will discuss social networks and their important role in entrepreneurship in detail in a later section. Overall, then, it is preferable for founding teams to be constructed on the basis of the principle of complementarity, because it adds to the total of the team's cognitive and social assets.

Hopefully, the major moral of this discussion is clear: when forming a founding team, entrepreneurs should do their best to avoid the seductive call of similarity. Yes, constructing teams entirely of persons similar to ourselves is tempting, and it makes working with them both comfortable and pleasant. But it is *not* the best strategy for making the founding team as strong and effective as it can be. The true "dream team" in this respect is one in which each member contributes something that the others do not provide, thus giving the old saying "the whole is greater than the sum of its parts" new and, for entrepreneurs, vital meaning.

Building social networks—and using them successfully

The preceding section emphasized the fact that a "Dream Team" of new venture founders would include people with non-overlapping social networks. It is now important to amplify that point because building and effectively utilizing social networks is of tremendous—perhaps central—importance to entrepreneurs (e.g., Aldrich and Kim, 2007). Here, even before we consider a formal definition of social networks, is a current example.

Note: In the popular television show, "Pawn Stars", Rick Harrison, owner of a famous pawn shop in Las Vegas, often calls upon help from members of his social network in determining the value of unusual or rare items such as this one—items that customers wish to sell. What is it worth? Mr. Harrison calls on help from appropriate experts to find out.

Source: Fotolia 15090456.

Figure 5.2 *One tangible benefit of social networks: help in valuing the rare*

Have you ever seen the television show *Pawn Stars*? In it, Rick Harrison and his family, who own a large pawn shop in Las Vegas, Nevada, negotiate deals with many different customers who bring an amazing array of items to the shop—everything from Super Bowl rings to a motorcycle owned by a famous movie star. In many cases, Harrison or other members of his family know the value of the items, but in others—for instance, an old and rare book, a special kind of antique gun—they do not. When they do not, they call on their network of experts to help. When these people appear, they provide an on-the-spot appraisal of the item in question, which then provides Steve Harrison, his father, or his son, with a starting point for negotiating with the customer. Clearly, the Harrisons have an extensive social network of experts—people who can provide them with estimates of a fair price for virtually anything, even objects they have never seen before (see Figure 5.2). And just as clearly, they could not run their business successfully without this help; without expert guidance they might pay too much for some items, fail to make an offer on others, and in general, be much less effective in operating their now famous shop.

The nature and impact of social networks—and why they are crucial for entrepreneurs

The example above (from *Pawn Stars*) illustrates what is, in essence, the heart of social networks. Formally, social networks are defined as a social structure composed of a set of members (individuals, organizations), and the ties between them, which indicate who knows whom, how well, and in what context(s). Social networks are important in many business settings, but are especially crucial for entrepreneurs because they provide the people in them with many valuable resources—information, legitimacy (knowing the "right" people can boost an unknown entrepreneur's status visibility), and influence (persons who appear to be legitimate are often better able to wield influence than those who lack this "certification"). In addition, they also provide psychological support for the persons in them, who are confident that they can get the help they need, when they need it from others in their networks (Lin, 1999).

More formally the term *social network* refers to a social structure composed of a set of members (which can be either individuals or organizations), and the ties between them—in essence, who knows whom, to what extent, and in what context. Social networks are important in many contexts, from sports to science, but they are especially crucial for entrepreneurs. Because they are attempting to create something new, and in many instances are founding a business that did not exist before, entrepreneurs have a great need for the benefits that social networks provide. Among these, information is certainly central. As noted above, founding teams—even the best ones—cannot possibly have all the information needed to successfully run every aspect of the new venture. For instance, perhaps the founding members are knowledgeable about producing new products, testing them, and marketing them; but they may know little or nothing about accounting, government rules concerning safety or the hiring of employees, or many other topics. An extensive social network can help provide such information.

Social capital: the benefits derived from social networks

When entrepreneurs receive financial support from venture capitalists, banks, or government agencies, these sources of financial resources also often help the entrepreneurs to build larger social networks. For instance, the best venture capital firms have huge social networks of their own, and so can put the entrepreneurs in touch with individuals or organizations who can help fill their needs, whatever these are. If, for instance, entrepreneurs need help with technical matters, the venture capitalists may know precisely the right sources of such help because they (the venture capitalists) often specialize in supporting only certain kinds of new ventures—ones in certain industries. Thus, the venture capitalists have ample opportunity to build social networks they can then put at the disposal of the entrepreneurs whose companies they choose to support.

In general, the benefits provided by a social network are described by the term *social capital*. Social capital has been defined in several different ways, but in general, refers to (1) the ability of individuals to obtain benefits from their social relationships with others, (2) these benefits themselves (Nahapiet and Ghoshal, 1998; Portes, 1998) or (3) the structure of individuals' social networks and their location in the larger social structure of the domain (e.g., industry) in which the entrepreneurs are functioning (see, e.g., Aarstad et al., 2009).

In brief, social capital refers to, and derives from, the social ties entrepreneurs have with others and the benefits they can obtain from these ties (Putnam, 2000). As you can probably guess, entrepreneurs with high levels of social capital (based on extensive and high-quality social networks) can readily obtain information and guidance when they need it; entrepreneurs lacking in social capital do not have such resources at their disposal.

In more general terms, social capital provides entrepreneurs with increased access to both tangible and intangible resources. Tangible benefits include financial resources and enhanced access to potentially valuable information. Intangible benefits include support, advice, and encouragement from others, as well as increased cooperation and trust from them. While the benefits provided by these latter (intangible) resources are somewhat difficult to measure in economic terms, they are often highly valuable to the persons who obtain them. Thus, social capital is an important asset for entrepreneurs—one that provides major benefits.

The social ties on which social capital rests exist within social networks (Aldrich and Kim, 2007), and are often divided into two major types: (1) close or strong ties—for example, the strong, intimate bonds that exist between members of a nuclear family, very close friends, of members of an entrepreneurial team, and (2) loose or weak ties—social linkages of the type that occur outside families or intimate friendships, among people who become acquainted but are not close friends or on intimate social terms with each other (e.g., Adler and Kwon, 2002; Putnam, 2000).

Close (strong) ties are often viewed as leading to, or at least being associated with what is known as *bonding social capital*—they generate relationships between individuals that are based on mutual trust. Establishing and maintaining such strong ties can be effortful, but once developed, they are a ready source of both tangible resources and information. Loose or weak ties, in contrast, lead to (or are associated with) bridging social capital—they are especially useful in providing individuals with information that might otherwise not be available to them, but since they do not involve high levels of mutual trust and confidence, are less likely to provide tangible resources. On the other hand, they require little effort to establish or maintain. Over time, loose (weak) ties sometimes develop into strong ones, in which case they would lead to relationships based on mutual trust.

Since social capital offers important benefits, and derives—to a large extent—from social networks, it is clear that one important ingredient in entrepreneurs' success is their capacity to build, and then reap benefits from, extensive, high-quality social networks. In fact, the social network perspective goes somewhat further: it suggests that an individual's position in a social network is by far the most important determinant of the information to which she or he has access. Persons with many weak ties have access to a broad range of information, while those with a few strong ties have access to a more restricted range of information—although they may gain access to closely held information (information that is not available for "public" consumption). Access to information, in turn, plays a key role in opportunity recognition. One view of this process is that it arises from information asymmetries—the possession, by some persons, of information that others do not have (e.g., Burt, 1992; Granovetter, 1973). So in a sense, the position an individual occupies in a social structure, and kind of ties she or he has with others, strongly determine whether she or he will identify opportunities and so engage in entrepreneurial activity. (This view is compatible with the pattern recognition approach described in Chapter 3, since the information provided by social networks can be invaluable in individuals' efforts to "connect the dots" between seemingly unrelated events or trends.)

These are eminently reasonable suggestions, and it seems clear that social networks, and the information and other resources they provide, play a key role in entrepreneurship. Social network theory, however, downplays or even largely ignores the role of the individual in these processes. In general, it assumes that an individual's position in a social structure is central and that personal skills or characteristics are relatively unimportant. Further, it does not consider in detail how individuals obtain these positions within social networks, and how they utilize their networks to obtain maximum benefit from them. In contrast, these issues are examined in detail in another perspective, one that has its origins in the field of psychology rather than the field of sociology, as does network theory—a perspective emphasizing the important role of social or political skills (e.g., Ferris et al., 2007), skills that assist individuals in getting along well with others, and so in forming positive relationships with them.

Beyond social capital: the importance of social/political effectiveness

Everyone belongs to social networks, but not all social networks are equal in size or quality, or in terms of the benefits they provide. Further, not all positions within them are equally desirable: some positions provide the individuals that hold them with ready access to a broad array of useful information and other valuable resources, while others offer a more restricted range of benefits. What determines the quality of an individual's social network and her or his position within it? One view is that an individual's political or social skills play a key role in these outcomes.

As you already know from your own experience, some people are much more popular and likable than others. Those high on these dimensions have been found to be ones who possess excellent social or political skills—an array of skills (all ones that can be learned or improved) that help them to get along well with others. Persons high in such skills tend to build social networks with very desirable attributes, and are also excellent at deriving maximum benefit from these networks. To examine the role of such skills in social networks, and in key aspects of entrepreneurship, we will first describe their basic nature, then examine their role in the development and use of social networks, and finally consider evidence that they are indeed truly an important ingredient in entrepreneurial excellence.

The nature of political and social skills

These two terms—social skills and political skills—are closely related, but derive from different research traditions. The term *social skills* was developed in the field of social psychology to refer to proficiencies that help people get along well with others (e.g., Kotsou et al., 2011). The term *political skills*, in contrast, derives from literature in several fields of management and refers to individuals' ability to effectively understand others at work and to use such knowledge to influence them to act in ways that enhance either personal or organizational objectives, or both (e.g., Ferris et al., 2007). Despite their different origins, though, social and political skills overlap to a great extent, so we will treat them as very similar, if not identical, here.

Both social and political skills have been found to play an important role in key organizational processes. To mention just a few of these effects, persons high in social/political skills, compared to persons low in such skills, are more successful as job candidates (e.g., Riggio and Throckmorton, 1988), receive higher performance reviews from supervisors (e.g., Robbins and DeNisi, 1994), attain faster promotions and higher salaries (e.g., Belliveau et al., 1995). Similarly, individuals high in social skills generally achieve greater success than do persons low in such skills in many different occupations (e.g., medicine, law, sales; Wayne et al., 1997), attain better results in negotiations (e.g., Lewicki et al., 2005), and often (although not always) achieve higher levels of task or job performance (e.g., Hochwarter et al., 2006). Political and social skills also exert strong effects on outcomes in many contexts outside the world of work. For example, persons high in various social skills tend to have wider social networks than do persons low in social skills (e.g., Diener and Seligman, 2002). Social and political skills have even been found to influence the outcomes of legal proceedings, with persons high in such skills attaining acquittals more often than persons low in such skills (e.g., Downs and Lyons, 1991). What, specifically, are these political and social skills? They take many different forms, but among the most important are these:

Social Astuteness (Social Perception): The capacity to perceive and understand others (their traits, feelings, intentions) accurately; persons high on this dimension recognize the subtleties of interpersonal interactions, a skill that helps them develop effective, positive relationships with others.

Interpersonal Influence: The ability to persuade others using an appropriate level of influence tactics (e.g., persuasion rather than threats).

Social adaptability: The ability to adapt to a wide range of social situations and to interact effectively with a wide range of persons.

Networking ability: The ability to form and expand social networks.

Expressiveness: The tendency to show one's emotions openly, in a form others can readily perceive.

Impression management (including apparent sincerity): The capacity to make a good first impression on others, through self-promotion (providing positive information about one's accomplishments or skills), ingratiation—actions designed to induce positive feelings or reactions in the target person, and the appearance of (plus, it can only be hoped, genuine) sincerity—meaning what one says and presenting oneself in an honest and accurate manner.

Political skills and the development—and utilization—of social networks

As noted earlier, social network theory does not generally focus on how individuals build social networks, obtain specific positions in them, and utilize them to obtain valuable resources. In contrast, the social and political skills perspective examines these issues in detail. Overall, it suggests that individuals (entrepreneurs) who are high in political and social skills develop social networks higher on three key dimensions than persons

Table 5.1 *The relative benefits—and costs—associated with strong and weak ties in social networks*

Action/Goal	*Strong Ties*	*Weak Ties*
Obtaining resources	Low effort High returns	High effort High returns
Obtaining information	Low effort Low returns	Low effort High returns

Note: As shown here, strong ties require much effort to establish and maintain, but can yield high returns in terms of tangible resources. In contrast, weak ties are easier to establish and maintain, but are less likely to provide tangible resources; however, they are very useful in terms of providing information. This mixed picture of costs and benefits is why entrepreneurs should seek to develop social networks consisting both of strong and weak ties.

Source: Prepared by Robert A. Baron.

lower in these skills. First, the networks they develop are higher on the dimension of *network composition*—the balance of strong and weak ties is better, thus providing better access to information (from weak ties) yet maintaining good access to resources (from strong ties).

Second, individuals high in political and social skills build networks that are higher in terms of *network efficiency*—they obtain a better ratio of effort expended in establishing and maintaining social relationships to the resource these networks provide. As shown in Table 5.1, establishing and maintaining close ties requires considerable time and effort, while establishing and maintaining weak ties does not. A network high in efficiency is one in which the entrepreneurs reap major benefits without expending undue effort; again, persons high in political and social skills are better at attaining this balance.

Third, persons high in political and social skill develop social networks higher on the dimension of *network effectiveness*. This refers to the quality of the interactions among network members. Networks high in effectiveness are ones in which members willingly exchange high-quality resources and information that benefits all members of the network.

Why do persons high in political and social skills obtain these benefits? Because they are simply better at establishing relationships with others and do so more easily—with less difficulty or effort. Further, the relationships they establish are more positive in nature, and are characterized by higher levels of trust and mutual understanding. Such networks, in turn, provide entrepreneurs with more and higher-quality resources—including information they can use to recognize or create new opportunities, obtain better advice and guidance, and a broader array of tangible resources with which to develop these opportunities. The result? They are higher in entrepreneurial excellence and more likely to attain the success they seek.

Do social and political skills play an important role in entrepreneurship? Evidence indicating that they do

Are social or political skills really important for entrepreneurs? Given their powerful impact in a wide range of business contexts (as described above), it seems reasonable to predict that this would be the case. Further, in a sense, they provide an extension of the benefits provided by social capital. Social capital, once acquired, does indeed provide many gains, including access to venture capitalists, potential customers, and potential employees. But once such access has been achieved—the door is "open"—something further must happen: the entrepreneurs must win the confidence, trust, and support of these people, so that the venture capitalists decide to offer the financial support the entrepreneurs seek, the customers place orders, and prospective employees sign on. In short, social capital does open doors, but once inside, outcomes depend strongly on the entrepreneurs' social and political skills—their effectiveness in interacting with others.

There are strong grounds for suggesting that such skills might be especially important for entrepreneurs. Often, they meet and interact with a wide range of persons, many of whom are strangers who are not currently in their social networks. Social and political skills can be very beneficial to them in such interactions, and in establishing relationships with these persons—in essence, making them part of their existing social networks.

Similarly, when presenting their "pitches" to venture capitalists, potential customers, and many other persons, entrepreneurs need excellent communication skills and high levels of persuasiveness; these, in turn, often rest, to a large extent, on social and political skills. For instance, being persuasive often requires accurate understanding of the reactions of one's audience (i.e., a high level of social astuteness). Entrepreneurs also frequently face the task of generating enthusiasm for their ideas, companies, or products in other persons; high levels of expressiveness may be helpful in this context. Finally, it is often essential for entrepreneurs to create good first impressions on others, since they and their companies may be relatively unknown, at least initially (i.e., they have low levels of social capital and little or no reputation).

In sum, there are several reasons why high levels of social or political skills may be especially important for entrepreneurs, and indeed, a growing body of empirical findings offer support for this suggestion. In an initial study of this issue (Baron and Markman, 2003), entrepreneurs working in two different industries (cosmetics, high-tech) completed a widely used and well-validated measure of social skills (e.g., Riggio, 1986). Entrepreneurs' scores on this measure were then related to one indicator of their financial success—the income these entrepreneurs earned from their new ventures over each of several years. Results indicated that several social skills (social

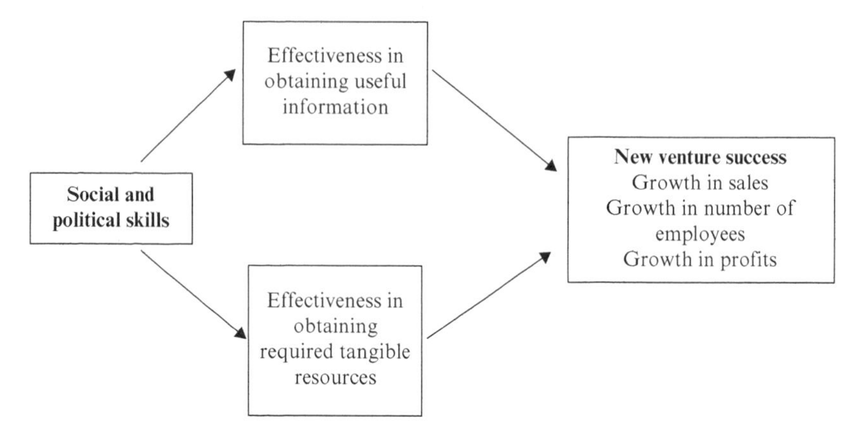

Note: Entrepreneurs who are high in social and political skills are more effective at obtaining useful information and essential resources than persons lower in such skills. These resources, in turn, contribute to the success of their new ventures.

Source: Based on findings reported by Baron and Tang, 2009.

Figure 5.3 *How entrepreneurs' political and social skills contribute to their success*

perception, social adaptability, expressiveness) were significantly related to this measure of financial success. Interestingly, when people who knew the entrepreneurs well rated their (i.e., the entrepreneurs') social skills, these ratings were highly correlated with the entrepreneurs' self-ratings. This suggests that people are relatively accurate in assessing their own social skills.

A follow-up investigation (Baron and Tang, 2009) extended these findings by investigating the underlying (i.e., mediating) mechanisms through which entrepreneurs' social skills influence the success of their new ventures. Results indicated that entrepreneurs' effectiveness in acquiring useful information and in obtaining essential resources both mediated the effects of their social or political skills on widely used measures of new venture performance, such as growth in sales, growth in profits, and growth in number of employees (e.g., Zahra et al., 2002). In other words, social or political skills helped entrepreneurs to obtain information and essential resources, and these resources, in turn, contributed to their success (see Figure 5.3). The study was conducted with Chinese entrepreneurs working in many different businesses; thus, it expanded earlier results to a very different cultural context and to many additional industries.

Fortunately, such skills can be readily acquired or strengthened (e.g., Kotsou et al., 2011). Decades of research by psychologists indicate that with careful, guided practice, most persons can increase these skills (e.g., Kurtz and Mueser, 2008). One technique that is often helpful is making recordings of individuals as they interact with others, and then having them review these recordings, while pointing out the shortcomings in their interaction style. Many people are surprised at their own actions, but can readily recognize the need for improvement in various ways. For instance, they may, inadvertently, be interrupting other persons, finishing their sentences for them—something many persons find irritating. When they view tapes of themselves doing this, persons who have developed this habit readily understand that it can be very annoying, and can learn to refrain from such actions.

Actually, many courses on entrepreneurship already provide this kind of feedback indirectly by requiring students to present their ideas for new products, services, or markets, to other members of the class and the faculty member in charge of the course. The feedback they receive may then include detailed comments on the style with which they present—not simply its content. While research findings (e.g., Chen et al., 2009) indicate that content is often more important (i.e., weighted more heavily in assessments of such presentations) than style—especially signs of excitement and passion about the ideas being presented (Cardon et al., 2009)—this latter aspect of a presentation matters too. Given the powerful impact of social and political skills on success in many business contexts, greater attention to this aspect might well prove highly beneficial for entrepreneurs-in-training.

At this point, we should note that research on both political skills and social networks combine to suggest specific actions current or potential entrepreneurs can use to build highly effective networks—ones that can provide the resources needed to build a successful new company. Here are some concrete suggestions (Lux et al., 2012).

- *Add "superconnectors"—people with rich social networks of their own—to your social network*: Such persons can be a tremendous help in building the large array of weak ties needed for maximum access to information.
- *Remember that quality often leads to quantity*: While social network size is important, even more important is the quality of the networks you develop. So eliminate relationships that are not positive or beneficial, and focus on establishing those that are.
- *Develop strong ties with resource providers but weak ties with information providers*: In general, only people with whom you have strong ties will provide you with concrete, tangible resources (financial or otherwise), so be sure to focus on developing strong ties with such persons. Ties with individuals who provide only information, however, can—and probably should be—weaker, which implies that there can be many more of them.
- *Consider relationship with others from both social and economic perspectives*: Some relationships with others are purely social—they are

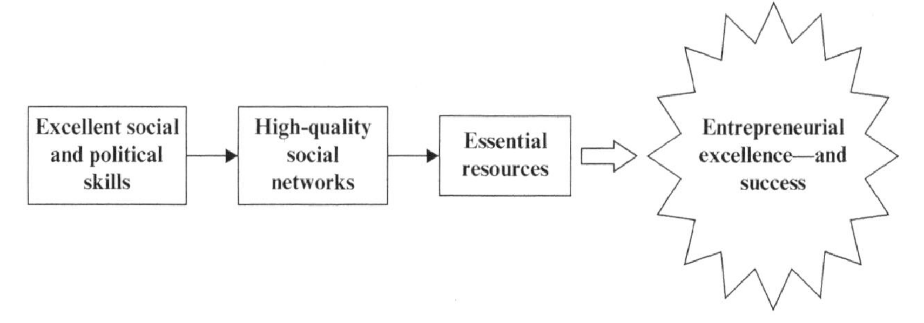

Note: Entrepreneurs who possess and effectively employ such skills tend to experience greater levels of success in starting new ventures than those who are lower in this respect, because they use their social and political skills to construct larger and higher-quality social networks. These networks then provide them with important resources that contribute to their success.

Source: Robert A. Baron.

Figure 5.4 *Social and political skills lead to high-quality social networks—and, ultimately, to entrepreneurial success*

enjoyable in and of themselves. Others, however, are established and maintained primarily because of the benefits they provide. It is important to recognize the differences between these relationships and, in most instances, to develop both kinds. The former will enhance the quality of your life, while the latter will increase the odds that you can obtain the resources needed for starting and developing a successful new venture.

To conclude: existing evidence indicates that in acquiring essential resources for their new companies or other entrepreneurial activities, entrepreneurs draw heavily on benefits conferred by their social networks, and the social capital provided by these networks. Further, their social and political skills appear to play a key role in building these networks. Overall, then, entrepreneurs who possess and effectively utilize such skills tend to experience greater levels of success in starting new ventures than those who are lower in this respect in part because they are able to construct larger and higher-quality social networks, and then can draw on these networks for important tangible and intangible benefits (see Figure 5.4). In a sense, therefore, social and political skills are a key component in the entire process: they help build the networks that provide the social capital that is such an important part of entrepreneurial excellence. So, are social and political skills important for entrepreneurs and should they seek to develop or strengthen them? Absolutely! For as Theodore Roosevelt once put it: "The most important single ingredient in the formula of success is knowing how to get along with people."

Summary of key points

Very few entrepreneurs attempt to develop their ideas alone; rather, most work in teams consisting of several people. Most of these teams are formed on the basis of similarity (homophily) between the members: they have similar backgrounds, training, interests, and even attitudes. Although similarity enhances the ease of communication, it also puts such teams at considerable risk for *groupthink* and other forces that can interfere with effective team performance. A much better way to form such teams is on the basis of complementarity, so that each member brings something unique and valuable to the team's knowledge, skills, and social contacts.

Social networks involve a social structure consisting of members (individual or organizations) and the ties between them. The many benefits individuals derive from membership in a social network include information and a wide array of both tangible (e.g., financial support) and intangible (e.g., psychological support) resources. Strong social ties (e.g., bonds between family members of close friends) are a major source of such resources, while weak social ties (e.g., those between casual business acquaintances) are an important source of information. Social and political skills possessed by individual entrepreneurs—skills that help them to get along well with others—also help them to build effective social networks. Such skills also help entrepreneurs to make their relationships with others high-quality ones, involving mutual trust and confidence. Important social and political skills include social astuteness, interpersonal influence, social adaptability, networking ability, expressiveness, and apparent (and hopefully actual) sincerity. Substantial evidence indicates that such skills are positively related to entrepreneurial success, so they should definitely be included in any list of factors contributing to entrepreneurial excellence.

References

Aarstad, J., Haugland, S.A., and Greve, A. (2009). Performance spillover effects in entrepreneurial networks: Assessing a dyadic theory of social capital. *Entrepreneurship Theory and Practice*, 34, 1003–1019.

Adler, P., and Kwon, S. (2002). Social capital: Prospects for a new concept. *Academy of Management Review*, 27, 17–40.

Aldrich, H.E., and Kim, P.H. (2007). Small worlds, infinite possibilities? How social networks affect entrepreneurial formation and search. *Strategic Entrepreneurship Journal*, 1, 1147–1166.

Bae, T.J. (2012). The role of advice-taking on venture performance. Paper presented at the Babson Entrepneuership Research Conference, Fort Worth, Texas.

Baron, R.A., and Markman, G.D. (2003). Beyond social capital: The role of entrepreneur's social competence in their financial success. *Journal of Business Venturing*, 18, 41–60.

Baron, R.A., and Tang, J. (2009). Entrepreneurs' social competence and new venture performance: Evidence on potential mediators and cross-industry generality. *Journal of Management*, 35, 282–306.

Belliveau, M.A., O'Reilly, C.A., III, and Wade, B. (1995). Social capital at the top: Effects of social similarity and status on CEO compensation. *Academy of Management Journal*, 39, 1568–1593.

Bledo, R., Rosing, K., and Frese, M. (in press, 2012). A dynamic perspective on affect and creativity. *Academy of Management Journal.*

Bonaccio, S., and Dalal, R.S. (2006). Advice taking and decision-making: An integrative literature review, and implications for the organizational sciences. *Organizational Behavior and Human Decision Processes*, 101, 127–151.

Burt, R. (1992). *Structural Holes: The social structure of competition.* Cambridge, MA: Harvard University Press.

Burt, R.S. (2000). The network entrepreneur. In Swedberg, R. (ed.), *Entrepreneurship: The social science view*, pp. 281–307. Oxford: Oxford University Press.

Byrne, D., and Griffitt, W. (1973). Interpersonal attraction. *Annual Review of Psychology*, 24, 317–336.

Cardon, M.S., Wincent, J., Singh, J., and Drnvosek, M. (2009). The nature and experience of entrepreneurial passion. *Academy of Management Review*, 34, 511–532.

Chen, X.P., Yao, X., and Kotha, S. (2009). Entrepreneur passion and preparedness in business plan presentations: A persuasion analysis of venture capitalists' funding decisions. *Academy of Management Journal*, 52, 199–214.

Diener, E., and Seligman, M.E.P. (2002). Very happy people. *Psychological Science*, 13, 81–84.

Downs, A.C., and Lyons, P.M. (1991). Natural observations of the links between attractiveness and initial legal judgments. *Personality and Social Psychology Bulletin*, 17, 541–547.

Ferris, G.R., Davidson, S.L., and Perrewé, P.L. (2005). *Political Skill at Work: Impact on work effectiveness*. Mountain View, CA: Davies-Black.

Ferris, G.R., Treadway, R.L., Perrewé, P.L., Brouer, D.C., Douglas, C., and Lux, S. (2007). Political skill in organizations. *Journal of Management*, 33, 290–320.

Ferris, G.R., Treadway, D.C., Kolodinsky, R.W., Hochwater, W.A., Kacmar, C.J., Douglas, C., and Frink, D.D. (2005). Development and validation of the political skill inventory. *Journal of Management*, 31, 126–152.

Granovetter, M.S. (1973). The strength of weak ties. *American Journal of Sociology*, 6, 1360–1380.

Hochwarter, W.A., Witt, L.A., Treadway, D.C., and Ferris, G.R. (2006). The interaction of social skill and organizational support on job performance. *Journal of Applied Psychology*, 91: 482–489.

Kotsou, I., Nelis, D., Gregoire, J., and Mikolajczak, M. (2011). Emotional plasticity: Conditions and effects of improving emotional competence in adulthood. *Journal of Applied Psychology*, 96, 827–839.

Kurtz, M.M., and Mueser, K.T. (2008). A meta-analysis of controlled search on social skills training for schizophrenia. *Journal of Consulting and Clinical Psychology*, 76, 291–304.

Lewicki, R.J., Saunders, D.M., and Barry, M. (2005). *Negotiation*. New York: McGraw Hill/Irwin.

Lin, N. (1999) Social networks and status attainment. *Annual Review of Sociology*, 25, 467–487.

Lux, S., Baron, R.A., and Ferris, G.R. (2012). The effects of social effectiveness on network development: The social networks of politically skilled individuals. Unpublished manuscript, University of South Florida.

Nahapiet, J., and Ghoshal, S. (1998). Social capital, intellectual capital, and the organizational advantage. *Academy of Management Review*, 23, 242–266.

Portes, A. (1998). Social capital. *Annual Review of Sociology*, 23, 1–24.

Putnam, F. (2000.) *Bowling Alone: The collapse and revival of American community*. New York: Simon & Schuster.

Riggio, R.E. (1986). Assessment of basic social skills. *Journal of Personality and Social Psychology*, 51, 649–660.

Riggio, R.E., and Throckmorton, B. (1988). The relative effects of verbal and nonverbal behavior, appearance, and social skills on valuations made in hiring interviews. *Journal of Applied Psychology*, 18, 331–348.

Robbins, T.L., and DeNisi, A.S. (1994). A closer look at interpersonal affect as a distinct influence on cognitive processing in performance evaluations. *Journal of Applied Psychology*, 79, 341–353.

Rueff, M., Aldrich, H.E., and Carter, N.M. (2003). The structure of founding teams: Homophily, strong ties, and isolation among U.S. entrepreneurs. *American Sociological Review*, 68, 195–222.

Treadway, D.C., Ferris, G.R., Duke, A.B., Adams, G.L., and Thatcher, J.B. (2007). The moderating role of subordinate political skill on supervisors' impressions of subordinate ingratiation and ratings of subordinate interpersonal facilitation. *Journal of Applied Psychology*, 92, 848–855.

Turner, M.E., and Pratkanis, A.R. (1998). Twenty-five years of groupthink theory and research: Lessons from the evaluation of a theory. *Organizational Behavior and Human Decision Processes*, 73, 105–155.

Wayne, S.J., Liden, R.C., Graf, I.K., and Ferris, G.R. (1997). The role of upward influence tactics in human resource decisions. *Personnel Psychology*, 50, 979–1006.

Zahra, S.A., Neubaum, D.O., and El-Hagrassey, G.M. (2002). Competitive analysis and new venture performance: Understanding the impact of strategic uncertainty and venture origin. *Entrepreneurship Theory and Practice*, 27, 1–28.

6 The personal side of entrepreneurial excellence: characteristics that enhance success

Chapter outline

Self-efficacy: confidence in being able to do the job—whatever it is!
Risk: are entrepreneurs really risk-takers? Should they be?
Key aspects of personality: do effective entrepreneurs share a personal style?
Emotion and passion: striking the right balance
The potential benefits—and costs—of being "up"
Passion: the intense desire to be—and act—entrepreneurially
The importance of being flexible: from business plans to improvisation and effectuation
Business plans and improvisation
Effectuation: using what you have to get where you want to go

* * *

> You can't ask customers what they want and then try to give that to them. By the time you get it built, they'll want something new.
> (Steve Jobs)

> I believe that one of life's greatest risks is never daring to risk.
> (Oprah Winfrey)

> If at first you don't succeed, try again. Then quit. There's no use being a damn fool about it.
> (W.C. Fields)

It has been several chapters and many pages, but perhaps you recall the discussion, in Chapter 1, of the question of whether entrepreneurs are a "breed apart"—different from most other persons? If so, perhaps you also recall that there are strong grounds for suggesting that they *are* different for very basic reasons: only some people are attracted to entrepreneurship, only some find that they are suited to it, and therefore, only some continue in this role (attraction–selection–attrition theory). So, there are compelling reasons for suggesting that entrepreneurs probably are different from most other persons in some respects, just as the members of any demanding, highly selected field or profession are different from persons in other fields (e.g., surgeons, philosophers, physicists, even politicians!). Saying that entrepreneurs are probably different is one thing; identifying the key ways in which they actually differ—and especially, the key dimensions related to their effectiveness as entrepreneurs—is another. Decades of research on individual differences indicate that people, and various groups, differ from one another in an almost countless number of ways. The key question, then, is which of these dimensions are most relevant for entrepreneurs? In this chapter, we will consider the ones for which there is strongest evidence—the personal characteristics and skills that appear to be most closely linked with entrepreneurial excellence. Once again, get ready for some surprises, because this evidence does not always fit closely with what "common sense" or informal observation suggests.

Self-efficacy: confidence in being able to do the job—whatever it is!

There is an old saying that, in one version or another, is popular in many different fields. In essence it says, "Confidence sells". What it means is that being confident—or perhaps simply having the appearance of confidence—actually increases the odds of success in many different contexts. As one well-known entrepreneur puts it: "The old saying that 'success breeds success' has something to it. It's that feeling of confidence that can banish negativity and procrastination and get you going the right way" (Donald Trump). Is this really true? In fact, research on success in many different fields and activities indicates that it is—the belief that we can accomplish whatever we set out to accomplish—is indeed a powerful factor in actually reaching our goals. A widely used term for this type of confidence is *self-efficacy* (Bandura, 1997, 2012), and it has been found to be one of the best predictors of success in everything from careers to building large social networks (see Chapter 5). Is this belief in their own ability to succeed also important for entrepreneurs? Given that they face uncertain situations lacking in clear guidelines or rules, and in which they must rely largely on their own skills and abilities, it seems reasonable to expect that self-efficacy might be especially important for entrepreneurs. If they did not believe in their own capacity to achieve what they set out to accomplish, they might, perhaps, never even begin!

A large body of evidence offers support for this prediction (Zhao et al., 2005). Overall, entrepreneurs are indeed higher in self-efficacy than other groups. Perhaps more directly, this evidence also indicates that there is a strong link between self-efficacy and both the tendency to start new ventures (Markman et al., 2002; Zhao et al., 2005) and to achieve success in them. So, just as is true in many other contexts, believing that one can "do it"—and do it well—is often an important plus for entrepreneurs. This is not to say that it is always beneficial, however. Entrepreneurs, as a group, are very high on optimism, and when high levels of self-confidence are combined with high levels of optimism (the belief, often unjustified, that everything will turn out well), the combination can be deadly: persons high on both dimensions may undertake tasks they are not qualified to perform and still expect to succeed at them (Hmieleski and Baron, 2009). In short, as is true for many other personal characteristics, there can be too much of a good thing where self-efficacy is concerned, especially if it is combined (as it often is among entrepreneurs) with high levels of optimism. The bottom-line conclusion, then, is this: Yes, confidence in one's abilities to perform well is generally a good thing—but as is true of many other characteristics, moderation—not excess—may be best (Grant and Schwartz, 2011).

Risk: are entrepreneurs really risk-takers? Should they be?

Entrepreneurs themselves (see the quotation by Oprah Winfrey above) often report that they are "risk-takers"—ready to take leaps into the unknown and into situations or activities with highly uncertain outcomes. Is this accurate? Recall that people are not very good at remembering either what they did or why they did it—memory is far from perfect and we often do not understand our own motives as well as we think we do (Chapter 3). Yet, early research findings seemed to confirm the view that entrepreneurs are in fact "high rollers" (e.g., Simon et al., 2000); they did appear to be more accepting of high levels of risk than other persons. (Risk-taking, by the way, is often defined as choosing options with high outcome variability—making choices that can result in a very wide range of possible outcomes; Figner and Weber, 2011.)

More recent evidence on the question of whether entrepreneurs are high risk-takers, however, offers a very mixed picture. In fact, two extensive and careful meta-analyses (statistical reviews of existing evidence; Miner and Raju, 2004; Stewart and Roth, 2001) reached sharply contrasting conclusions concerning risk-taking by entrepreneurs. One (Stewart and Roth, 2001), concluded that entrepreneurs are indeed more accepting of risk than other persons, while the second (Miner and Raju, 2004) reached precisely the opposite conclusion. While many differences between the methods used in the two reviews may account for these contrasting conclusions, another explanation is as follows: perhaps the propensity to accept risk changes over different phases of the entrepreneurial process. Early (when a new venture or other entrepreneurial activity is just getting started or perhaps has not yet begun), acceptance of relatively high levels of risk is virtually required; individuals with a strong aversion to risk would never begin the process. With failure (or at least, closing down) rates for new ventures in the range of 80–85 percent during the first three years, it seems clear that the attraction–selection–attrition model would operate—and very strongly—so that only persons willing to take risks would become, and remain, entrepreneurs (e.g., Ariely, 2009).

On the other hand, during later phases of the process, once a new venture has been launched or other entrepreneurial action is well under way, the goal of protecting and stretching existing resources might become dominant, with the result that entrepreneurs (and certainly successful ones) strive actively to limit or manage risk. So, in answer to the question, "Are entrepreneurs really risk-takers?" the answer may depend (1) on how we measure risk, and (2) when, during the entrepreneurial process, it is measured.

Several other factors, too, may play a role, and make it difficult to answer the question "Are entrepreneurs prone to taking risks?" in a simple or straightforward way. First, entrepreneurs may not differ from others in terms of willingness to accept risk, but rather in their perceptions of risk. That is, in a given situation, entrepreneurs tend to perceive risks as lower than other people do (Busenitz and Barney, 1997). Why this should be so, is unclear, but perhaps being high in optimism, they perceive the likelihood of positive outcomes as higher than it actually is. Another complicating factor is that risk-taking is often strongly influenced by the situations in which it occurs. For instance, people differ in their willingness to take risks with respect to different activities or aspects of their lives—when gambling (e.g., betting on a horse race or playing the lottery), investing, engaging in actions that may influence their own health and safety, and in situations involving ethical decisions or actions (Figner et al., 2009). For instance, an individual who is perfectly willing to take risks when investing (e.g., willing to buy the shares of new companies that have little or no track record of earnings), may be much less willing to take risks with her or his personal health or to make decisions with strong ethical implications, such as claiming unjustified deductions on her or his income tax returns. Finally, risk-taking may also be influenced by other personal characteristics, aside from whether someone is an entrepreneur—for instance, age; in general, willingness to take risks declines from the teen-years through the later decades of life; and optimism, with people high in this characteristic are more willing to take risks than ones low in optimism (see Figure 6.1 for a summary of these suggestions concerning the factors that influence risk-taking).

So, are entrepreneurs risk-takers? In a general sense, the answer is probably "no"—overall, they are not more willing to accept high levels of risk than other persons. On the other hand, they may be more willing to accept risk with respect in certain situations—for instance, in investing

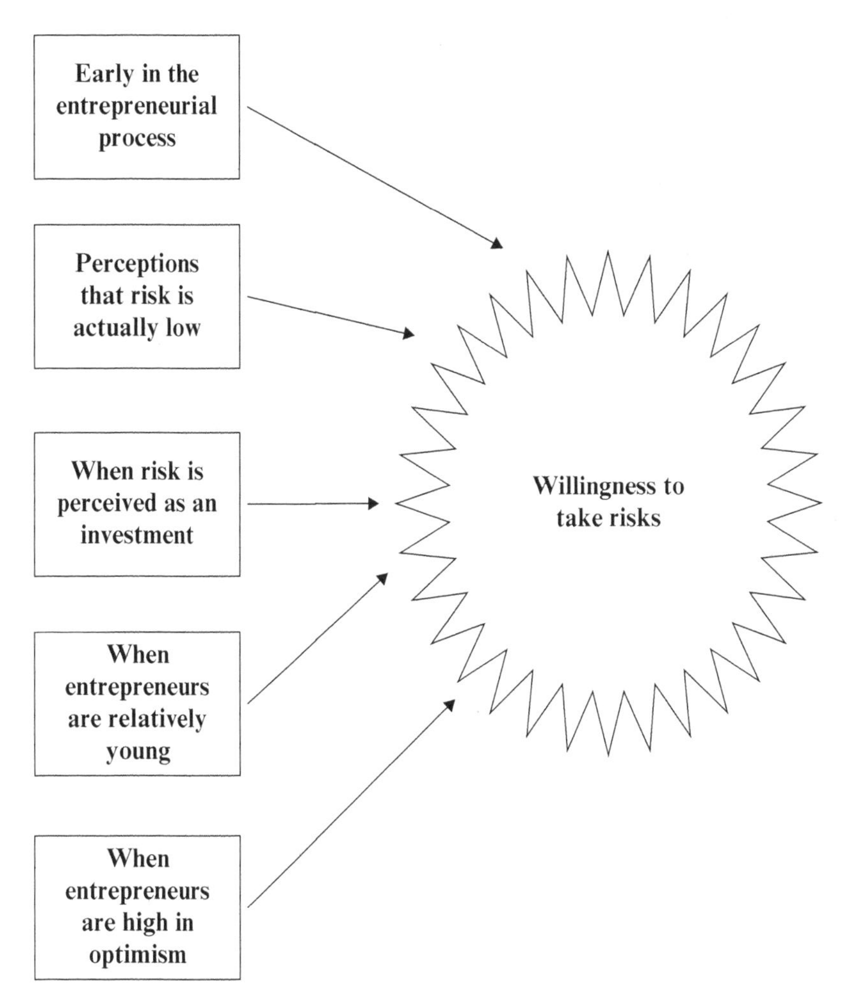

Note: Research on the question of whether entrepreneurs are high risk-takers is mixed, but suggests that overall, entrepreneurs are not necessarily willing to accept high levels of risk. Rather, they are more willing to take risks under some conditions than others.

Source: Robert A. Baron.

Figure 6.1 *Are entrepreneurs high risk-takers? Only under some conditions*

their own time, efforts, and resources in their new ventures; and some recent findings (e.g., Baron et al., under review, 2012), indicate that they be more willing to "cut corners" with respect to some kinds of ethical decisions, especially when they believe that doing so is essential to keeping their new ventures alive. The general belief that entrepreneurs are "high rollers" where risk is concerned, however, seems to represent an instance in which informal, common-sense-based knowledge provides a somewhat incomplete and inaccurate answer to an important question.

Key aspects of personality: do effective entrepreneurs share a personal style?

At one time, the term *personality* had a very negative connotation in the field of entrepreneurship. To some extent, this stemmed from a misunderstanding about the meaning of this term. Basically, it simply refers to predispositions or preferences—what individuals would prefer to do when they are free to think, feel, and act the way they wish (Larson and Buss, 2009). It certainly does not refer to traits or attributes that are "set in stone"—ones that are unchanging over many years (perhaps an entire lifetime) and in fact, cannot readily be altered. Although personality is stable, it also shows considerable change as people live their lives. In any case, for many years, it was widely believed that "personality" was such a complex and poorly understood concept, that it could not usefully be applied to the task of understanding entrepreneurial excellence.

This situation changed when attention was shifted to a framework for understanding personality that was developed in the 1990s, known as "the big five dimensions" of personality (Barrick and Mount, 1991). Literally hundreds of separate studies conducted in several different fields suggest that this model does represent what appear to be very basic aspects of individual behavior—relatively stable differences in their preferences, dispositions or—if you prefer—personal style. Here are brief descriptions of the high ends of each of these dimensions. The low ends, of course, are opposite in nature. (These are indeed dimensions because individuals vary along each from very low to very high.)

1. *Conscientiousness*—the tendency to be high in achievement and work motivation, organization and planning, self-control, responsibility.
2. *Openness to experience*—the tendency to be high in curiosity, imagination, creativity, seeking out new ideas.
3. *Extraversion*—the tendency to be outgoing, warm, friendly, and energetic.
4. *Agreeableness*—the tendency to be trusting, cooperative, and altruistic in one's dealing with others.
5. *Emotional stability*—tendencies to be calm, stable, even-tempered, and hardy (i.e., resilient) in the face of high levels of stress.

These dimensions have been shown to be related to a wide range of important outcomes, ranging from job performance (Barrick and Mount, 1991) to the size and quality of individuals' social networks. That they do indeed reflect very basic aspects of human behavior is indicated by research such as this: it is arranged for strangers to meet for the first time. After spending a few minutes together, they rate each other on the "Big

Five Dimensions". When these ratings are compared with those provided by persons who know the research participants very well (e.g., close relatives, spouses, co-workers), the two sets of ratings correlate highly (Mount et al., 1994). This suggests that where people stand on several of these dimensions is readily apparent to people meeting them for the very first time. This may be one reason why "speed dating" works: participants can form relatively accurate impressions of key aspects of each other's personal style in just a few minutes (Baron and Branscombe, 2012). Such impressions are far from totally reliable (as we saw in Chapter 5, many people can "manage" their behavior in order to make a good impression on others), but they do provide a good starting point for initiating the process.

Much more important for this discussion, research findings indicate that several of these dimensions—especially conscientiousness, openness to experience, and emotional stability—are significantly linked to becoming an entrepreneur (Zhao and Siebert, 2006), and to entrepreneurs' success in this role. Entrepreneurs are higher than other groups (e.g., managers) on all three of these dimensions. However, findings are more mixed with respect to extraversion: although some results indicate that entrepreneurs are higher in extraversion than other persons (e.g., Zhao et al., 2010) other research results do not confirm this relationship. Finally, there is some indication that agreeableness is actually negatively related to becoming an entrepreneur—a finding indicating that in order to develop something new and better, it may often be necessary to be somewhat cautious or even self-centered in one's dealing with others—especially people with whom one is linked by weak ties (e.g., casual business acquaintances).

Perhaps even more important, these basic dimensions of personality are also related to entrepreneurs' success—for instance, the survival of their new ventures. Research findings indicate that conscientiousness is positively related to firm survival and that perhaps extraversion and openness to experience are also linked to entrepreneurial excellence—and success (Ciavarella et al., 2004). Overall, then, we can offer the following general conclusion: certain personal tendencies or dispositions are in fact helpful to entrepreneurs in their efforts to convert the ideas generated by their personal creativity into something new and better than what currently exists.

Emotion and passion: striking the right balance

Have you ever seen a *Star Trek* movie, or even the original TV show? If so, you know that one of the characters in it—Mr. Spock, the Chief Science Officer of the Starship Enterprise (a name assigned to one of the U.S. shuttles!)—was, supposedly, devoid of all emotion. He had no moods or feelings, no anger, no sorrow, no joy. Of course, the plots of these films often revealed that he *did* have these reactions—he did possess a "feeling side" to his existence as well as a purely cognitive one. That is far from surprising, because, it is hard to imagine a human being (or other intelligent being!) totally lacking in emotions and moods. This emotional or "feeling" side of life is very basic, and may pre-date cognition in an evolutionary sense. Structures and systems within our brains that appear to be intimately linked to emotions are ones we share with many other species, while structures and systems that are more central to what are known as *higher mental processes* (reasoning, decision-making, planning, etc.) are ones that are either uniquely human or that we share only with a smaller number of species. Further, there is a large body of evidence indicating that our feelings and emotions interact with various aspects of cognition so that, for instance, emotions and moods influence our decisions, planning, and reasoning (Baron and Branscombe, 2012). This has important implications for entrepreneurs for several reasons.

First, entrepreneurs clearly have a very rich and varied emotional life—perhaps more varied and intense than persons holding jobs in well-established organizations (Baron, 2008). The joys of success are greater, and the depths of despair deeper because their lives are less predictable, and their outcomes far less certain than is true for many other persons. Further, although emotions are experienced by individuals, growing evidence indicates that they permeate groups as well, and that group affect—affective states (feelings and moods) arising from interaction between group members—exist and can exert strong effects on the group's functioning and performance (e.g., Barsade and Gibson, 2011). In fact, group affect—the average level of affect experienced by group members—has been found to influence a wide range of actions, including willingness to help others (e.g., customers) and participate constructively in group activities, to withdrawal from them, and performance of many challenging tasks (e.g., George, 1990; Knight, 2011). Since most new ventures—and a large proportion of entrepreneurial activities—involve teams of several persons, and since feelings experienced by one member are often transmitted to others (a process known as *emotional contagion*; Barsade, 2002)—emotions do not influence merely the behavior of individuals, but that of groups or teams as well.

This raises a basic question: if the "feeling" side of life is so important, what are its effects on entrepreneurs and their new ventures? Research on this topic has investigated many issues, but here, we will focus on two that are, perhaps, most central to entrepreneurship: the potential benefits and costs of positive affect (experiencing positive moods and emotions often and in many situations), and the role of *passion* in entrepreneurship—the role of powerful identification with and the intense desires to participate in, entrepreneurial activities (e.g., Cardon, et al., 2009; Murnieks et al., in press, 2012).

The potential benefits—and costs—of being "up"

There is a strong general belief that being "positive"—experiencing and expressing positive feelings—is an important plus in life. And in fact, decades of research on the effects of positive affect—experiencing moods and emotions—suggests that this is generally true: the tendency to experience and express positive affect is strongly associated with many desirable outcomes (Ashby et al., 1999; Kaplan et al., 2009; Lyubomirsky et al., 2005; Weiss and Cropanzano, 1996) including: (1) increased energy, (2) enhanced cognitive flexibility, (3) increased creativity, (4) greater confidence and self-efficacy, (5) adoption of efficient decision-making strategies, (6) increased use of effective techniques for solving problems, (7) improved ability to cope with stress and adversity (Ashby et al., 1999; Baron, 2008; Fredrickson, 2001; see Chapter 8). In addition, and more generally, high levels of positive affect have been found to be related to improved performance on a wide range of cognitive and work-related tasks (Kaplan et al., 2009), increased career success, formation of more extensive and higher quality personal relationships (Baas et al., 2008; Lyubomirsky et al., 2005) and even enhanced personal health, both physical and psychological.

Given the extent of these effects, there appear to be strong grounds for concluding that positive affect either produces, or is associated with, a wide range of beneficial effects. The overall picture provided by research evidence is not entirely consistent, however. Although most evidence suggests that the effects of positive affect are indeed beneficial some findings are inconsistent with this overall pattern (Judge and Ilies, 2004). For example, previous research has reported that high levels of positive affect increase susceptibility to cognitive errors that can potentially interfere with effective decision-making (Isen, 2000), and can reduce performance on many tasks, especially ones involving critical reasoning and logic (Melton, 1995). Similarly, research by Zhou and George (2007) indicates that high levels of positive affect may not always facilitate creativity. And recent findings indicate that although positive affect generally increases performance on many tasks, if negative affect is followed by positive affect, it, too, can enhance performance (Bledow et al., 2011). In addition, high levels of positive affect have been found to reduce attention to negative information—especially, information that contradicts currently held beliefs and attitudes (Forgas and George, 2001). Clearly, ignoring negative input can be a very dangerous tendency for entrepreneurs. Finally, high levels of positive affect—especially forms of positive affect that are high in both positive valence and activation (e.g., enthusiasm, excitement)

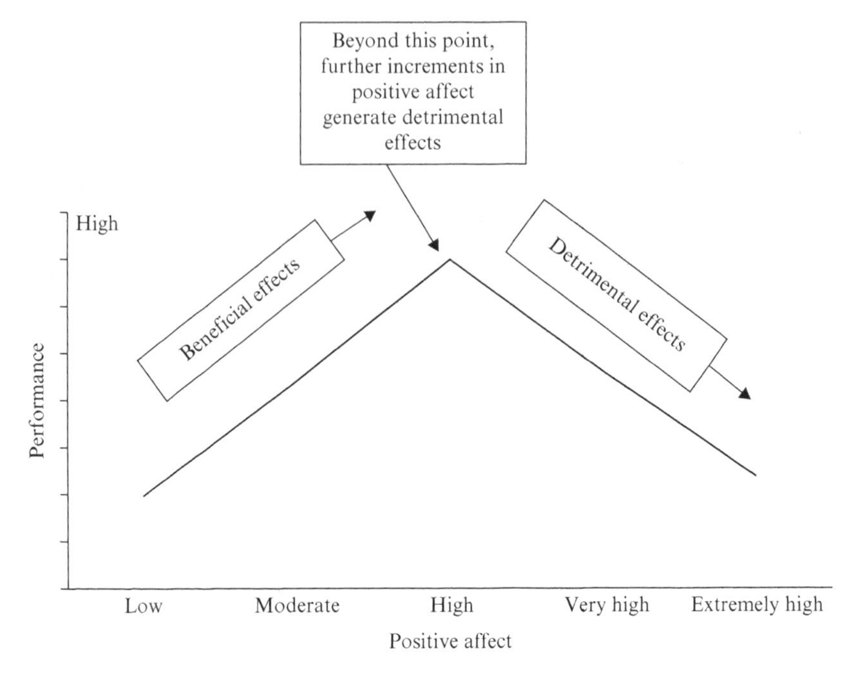

Note: As shown here, positive affect—the tendency to experience positive moods and emotions frequently and across situations—is associated with many beneficial effects, but only up to a point: beyond some level, it may actually have detrimental effects on cognition, behavior, and performance.

Source: Robert A. Baron.

Figure 6.2 *The benefits—and potential costs—of high positive affect (being "up")*

have been found to encourage *impulsiveness*—the tendency to act without adequate thought, abruptly, and with little or no regard for potential negative consequences (DeYoung, 2010). As we saw in Chapter 5, that, too, can be highly detrimental for entrepreneurs, whose new ventures generally have limited resources and cannot easily recover from the detrimental effects of rash actions or hasty decisions by their founders (Khaire, 2010).

How can "being up" (experiencing positive affect) have both beneficial and detrimental effects on many aspects of life? The answer that has emerged from recent research is that the relationship between positive affect and many outcomes is curvilinear, not linear in nature. Up to a specific point, increases in positive affect are associated with beneficial outcomes, but beyond this level, they become "too much of a good thing" and can actually interfere with important aspects of cognition and behavior (Grant and Schwartz, 2011; see Figure 6.2).

This suggestion is consistent not only with empirical evidence, but also several theories concerning the influence of affect on cognition and behavior (e.g., optimum level of affect theory; Oishi et al., 2007;

broaden-and-build theory; Fredrickson and Losada, 2005). These theories converge in suggesting that there may be limits to the beneficial effects of positive affect so that, overall, the relationship between such affect and performance on many different tasks takes the form of an inverted letter U, as shown in Figure 6.2.

Given that entrepreneurs tend, as a group, to be very high in positive affect (higher, in fact, than any other tested group; Baron et al., 2012), this fact has important implications for entrepreneurship. First, very high levels of positive affect (levels that are in a sense, too high?) may be one of many factors that interferes with entrepreneurs' success. Some evidence for this conclusion is provided by a recent study by Baron et al. (2011) who found that up to a discrete level, entrepreneurs' positive affect was positively related to their new venture's growth in sales and innovativeness but that beyond that point, actual declines in these measures occurred. In sum, entrepreneurs' high levels of enthusiasm, confidence, and optimism may serve both as a source of personal strength *and* a source of potential danger. Unless they can restrain their own tendencies to be "upbeat", they run the risk that their strong proclivity to experience positive affect can interfere with their own effectiveness. A key task for entrepreneurs, then, is that of learning to effectively restrain or managing these feelings. As we noted in Chapter 4, the capacity to regulate their own emotions—to keep these "in check" if you will—is an important self-regulatory skill for entrepreneurs. In a sense, this discussion expands and re-emphasizes the earlier discussion of this issue. In sum, entrepreneurs, if they wish to succeed, must be wary of the trap of being *too* positive. Perhaps Grant and Schwartz (2011, p. 62) put it best when they note: "there is no such thing as an unmitigated good. All positive traits, states, and experiences have costs that at high levels may begin to outweigh their benefits." That is an important and valuable warning for entrepreneurs to heed as they engage in efforts to make their ideas, dreams, and visions real.

Passion: the intense desire to be—and act—entrepreneurially

In commenting on his own success, Albert Einstein once said: "I have no special talents. I am only passionately curious." History suggests that he was being too modest—far too modest! But he was also being insightful, because over the centuries, many thoughtful observers have suggested that a crucial key to success is *passion*—intense involvement in and love for whatever one does. This seems especially to be true in creative activities, whether building theories of the universe, generating art, literature, or—in the case of entrepreneurs—ideas for something new that can significantly change the lives of large numbers of persons. Does passion also play a role in entrepreneurship? As an activity with creativity at its very core, it

Table 6.1 *Measuring entrepreneurial passion*

Aspect of Passion	*Item*
Positive feelings	Searching for new opportunities is very exciting for me.
	I greatly enjoy figuring out how to improve existing products of services.
Inventing	Inventing new ways to solve problems is important for me as an individual.
	Looking for new opportunities gets me very excited.
Developing	Developing companies and helping them grow is important to me as a person.
	Building a team of excellent people for my new company is very enjoyable.
Founding	Starting a new company is something I greatly enjoy.
	Being the founder of a company is an important part of my self-identity.

Note: The items shown here are similar to those used to measure various aspects of entrepreneurial passion. In the actual scale, individual items are designed to assess combinations of these factors—for example, positive feelings coupled with a strong desire to invent something new.

Source: Similar to items in Cardon et al., 2012.

seems likely that this is so, and in fact, many entrepreneurs have referred to the importance of passion in their work. But what, precisely, is passion? Psychologists (e.g., Vallerand et al., 2003) define it as a strong inclination toward an activity people like and find important. In the field of entrepreneurship, Cardon et al. (2009) and Murnieks et al. (in press, 2012) define it as a strong, positive inclination toward entrepreneurial activities, coupled with a powerful identification with these activities. In other words, entrepreneurs high in passion love what they do, and identify strongly with the role of being an entrepreneur.

Can such passion be measured? Cardon et al. (in press, 2012) have developed a brief questionnaire that accurately reliably assesses entrepreneurial passion. Table 6.1 shows some items similar to those on the scale, and as you can see, it is designed to measure several aspects of passion: intense positive feelings about entrepreneurial activities (e.g., "I strongly enjoy trying to figure out how to make existing products better"), the extent to which being an entrepreneur is central to the individual's identity (e.g., "Owning my own company very attractive to me") and identification with various roles that entrepreneurs play—inventor, founder, developer.

A key question, of course, is this: Are scores on this scale—which measures entrepreneurial passion—related to what entrepreneurs actually do and the success they achieve? Growing evidence suggests that in both cases, the answer is "Yes". Passion, as measured by the scale mentioned above, is related, first, to entrepreneurial self-efficacy—the belief among entrepreneurs that they can accomplish tasks related to their entrepreneurial activities (see earlier discussion in this chapter). In a sense, that is far from surprising, since entrepreneurs intensely enjoy these activities, and positive affect tends to enhance self-efficacy.

Perhaps even more important, passion has also been found to be related to entrepreneurial behavior, measured in terms of the amount of their available time they spend being as entrepreneurs rather than engaging in other activities (Murnieks et al., in press, 2012). Existing research does not yet indicate whether passion is also related to entrepreneurial success, but given the nature of passion—which involves strong positive affect plus powerful identification with being an entrepreneur and engaging in activities central to this role—it seems likely that this relationship, too, will be confirmed.

It is important to note, though, that although most research suggests that passion may be beneficial for entrepreneurs in several ways, at least one recent study indicates that entrepreneurial passion may not enhance one aspect of entrepreneurs' performance: their success in obtaining support (financial, human) for their ideas. Since entrepreneurial passion involves high levels of enthusiasm and enthusiasm often does "sell", it seems reasonable to expect that this would be the case. However, Chen et al. (2009) have reported that in making their "pitches" to venture capitalists and others, entrepreneurs' preparation may be more important than outward signs of their passion. Specifically, Chen et al. (2009) had professional actors make such "pitches". In these appeals for support, the actors showed either high or low levels or preparation (the extent to which the arguments presented were strong and convincing), and either high or low levels of passion (as evidence in nonverbal cues such as facial expressions and enthusiasm). After watching one of these presentations, participants (MBA students in one study and actual venture capitalists in another) indicate whether they would or would not make an investment in the new company. Results in both studies (with MBA students and actual venture capitalists) yielded the same results: preparedness seemed to be more important than passion in these decisions.

Does this mean that passion does not "pay" in terms of obtaining needed financial resources? Perhaps; but it may also be that passion is more important in other ways. For instance, it may help entrepreneurs maintain their commitment even in the face of major setbacks (a topic we will discuss in Chapter 8). Further, it is possible that this is yet one more instance of the "too much of a good thing" phenomenon described earlier. Up to a point, outward signs of passion may indeed help persuade venture capitalists and others to offer support for a new venture. However, beyond some point (and that point may have been reached or exceeded by the professional actors in the research by Chen et al.), extremely high levels of passion may be perceived as signs of insincerity or as efforts to introduce irrelevant and distracting elements into the situation (e.g., by attempting to dazzle the audience with presentations that are so enthusiastic in nature that it loses sight of important business considerations). Pending the completion of further research, no firm conclusions are yet possible. But the weight of existing evidence does seem to be tipping in favor of the suggestion that overall, passion is beneficial for entrepreneurs, and may be one more factor that plays a significant role in their success.

The importance of being flexible: from business plans to improvisation and effectuation

In most university programs focused on entrepreneurship, students must—at some point in their studies—prepare a formal business plan. This is a document in which entrepreneurs (current or would-be), describe, in detail, how they plan to accomplish the process that is the focus of this entire book: convert their ideas into reality—functioning companies that will generate profits or other benefits (e.g., contribute to the welfare of their communities). This project is often viewed as the "capstone" of many courses or even the entire training program: it counts for a large proportion of course grades, and is considered to be the ultimate demonstration, by students, that they have mastered the basic requirements for being an entrepreneur. Further, business plans are often the "entry card" into competitions for gaining financial and other resources. Teams of students compete in spirited business plan competitions whose winners gain not merely prestige; they also often receive large cash prizes, too (e.g., $25,000 or more is not uncommon for first place). In addition, the victors in such competitions get to meet venture capitalists, angel investors, and others who are capable of helping them realize their dreams. So yes, writing excellent business plans is serious business in programs, departments, and schools of entrepreneurship.

Business plans and improvisation

The basic rationale behind this focus is clear: detailed planning is usually necessary to developing successful companies. In other words, it is not enough to have clear goals—it is also important to have a plan for reaching them. And that is what business plans are all about. In addition, writing a business plan provides other advantages. As one writer (who was also a famous aviator in the period when airplanes were new) put it: "A goal without a plan is just a wish" (Antoine de Saint-Exuprey). Such planning, in turn, provides several benefits. First, an excellent business plan helps entrepreneurs sharpen their thinking about the opportunities they wish to pursue. Writing such a plan also helps them to get a clearer picture of the many complex issues they must resolve, and this, in turn, often helps them to formulate better and clearer ideas about how to proceed.

Second, a comprehensive business plan is, as suggested above, a kind of "selling tool"—it provides concrete information, presented in a standard format that potential sources of financial, human, or technical assistance can examine. This is how venture capitalists often proceed: first, they read business plans (or at least the executive summary, which provides an overview of the entire document), and then decide which teams to invite for interviews and presentations. So a business plan is indeed essential in cases where entrepreneurs need, and seek, external help. (In situations where they do not, however, this "selling tool" function may be less crucial.)

What does a comprehensive business plan contain? Basically, it describes the opportunity (e.g., what needs the new product or service will meet, what markets it will have), how the product or service will be produced or delivered, what skills and abilities the founding team brings to the new venture, how the new company will be structured, how the new venture will gain a competitive advantage, what critical risks it faces, what financial resources it needs, and how the products or services will be marketed. It usually often contains financial projections, but given that most entrepreneurs are highly enthusiastic about their own ideas, these are often viewed with healthy skepticism because most novice entrepreneurs are far too rosy in their predictions!

In a sense, then, "business plans" reign supreme in the academic field of entrepreneurship: teaching students to prepare them is one of the key goals sought in such programs. Is this appropriate? Should so much emphasis be placed in this kind of detailed planning? Actually, there is growing controversy about this issue, and for several compelling reasons. First, as you can guess, preparing a detailed business plan requires a great deal of effort—that, in itself, is not at all surprising. The surprise, however, resides in the fact that there is, at present, very little compelling evidence that preparing such plans contributes significantly to the success of new ventures! In fact, existing evidence on this issue is mixed. Some studies indicate that preparing detailed written business plans is related to certain aspects of new venture success—for instance, the amount of financial resources they raise (e.g., Lange et al., 2007). Other investigations, however, find no clear link between preparing a formal business plan and subsequent success (Brinckmann et al., 2010). In these studies, preparing a business plan was not significantly related to standard measures of success, such as growth in sales, earnings, profits, or return on investment.

Together, this mixed picture of results points to the possibility that while planning is often beneficial, even here, there can be too much of a good thing. And anyone who has ever been an entrepreneur (including the present author) understands almost intuitively why this may be so. Once a new venture is launched, or other entrepreneurial activity begins, very little, if anything, goes according to plan. In fact, in most instances, entrepreneurs soon find that they must depart from their business plans in many ways—often very major ones. This drives home a point made at several points in this book: As they proceed, entrepreneurs do indeed often have to "make it up as they go along". In short, they must engage in improvisation, which means performing extemporaneous actions—ones that occur without formal planning (Hmieleski and Corbett, 2008).

The necessity to improvise is generated by the fact that the conditions entrepreneurs face are ever-changing. Competitors offer new products or improvements on existing ones, government policies and regulations change, new markets open while others shrink in size, technology presents new opportunities—and eliminates others, and so on. Under such fluid and unpredictable conditions, entrepreneurs simply *must* improvise if they are to succeed, and research on the benefits of improvisation indicate that it is indeed highly beneficial (e.g., Mullins and Komisar, 2009).

In fact, current evidence indicates that entrepreneurs willing and able to improvise tend to be more successful than ones lower on this dimension (e.g., Hmieleski and Corbett, 2008). These benefits are greater under some conditions than others, however. For example, willingness to improvise is more strongly related to success (high levels of new ventures' performance) for entrepreneurs high in self-efficacy than it is for entrepreneurs lower in this characteristic (Hmieleski and Corbett, 2008). This is because entrepreneurs who improvise are, in a sense, taking a leap into the unknown and are unlikely to hit on the best formula for attaining success immediately. Rather, they often need to adjust and re-adjust their plans and strategies, and persons high in self-efficacy are better prepared to persevere through this difficult process of trial-and-error (Bandura, 2012). Other findings indicate that willingness to improvise is more beneficial to entrepreneurs moderate rather than high in optimism, and ones who work in highly dynamic (i.e., rapidly changing) rather than stable environments (Hmieleski et al., in press, 2012). In fact, these conditions operate together to influence firm performance. In dynamic environments (ones that change rapidly) the tendency to improvise is positively related to new venture success, but primarily for entrepreneurs who are moderate rather than very high in optimism. This is because high levels of optimism seem to interfere with the tendency to make continuous changes and adjustments; persons very high in optimism believe that everything will turn out well, so they tend to stick with the courses of action they choose initially. In stable environments, however, entrepreneurs' tendency to improvise is significantly related to the performance of their companies, regardless of whether they

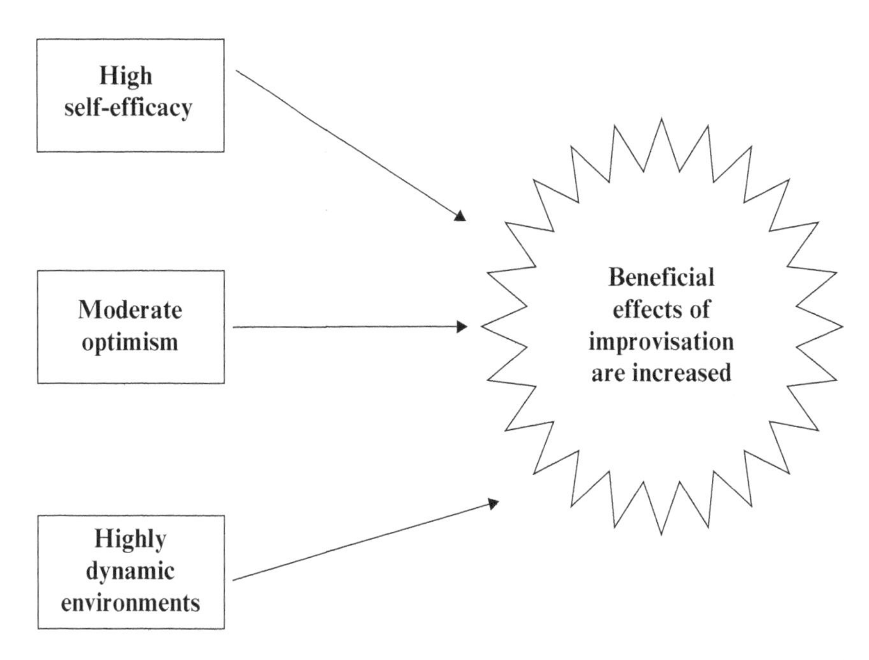

Note: Being willing and able to improvise is generally beneficial for entrepreneurs, but it is especially helpful when they are high in self-efficacy, moderate in optimism, and face rapidly changing (highly dynamic) environments.

Source: Based on findings reported by Hmieleski et al., 2012.

Figure 6.3 *Willingness to improvise: An important "plus" for entrepreneurs*

are moderate or high in optimism. In sum, the tendency to improvise can indeed be beneficial, but not always and not under all conditions. As shown Figure 6.3, it is especially beneficial for entrepreneurs high in self-efficacy, and moderate in optimism, who face rapidly changing conditions.

Effectuation: using what you have to get where you want to go

The winter after my family moved to Oklahoma (from upstate New York), a major snow storm occurred in our area. This is very rare in central Oklahoma, so I had sold or given away all of my snow-removal equipment before moving. Now, I suddenly found that I desperately needed to clear my driveway so I could get to work. What could I do? I looked around the house and garage, and found a solution: I took a broom handle and attached a plastic dish draining board to it. The result? A home-made snow shovel that did the trick; in fact, it lasted through another storm as well before the plastic cracked and it became unusable.

This personal experience, trivial as it is, provides an illustration of a process known as *effectuation*—one that has recently received growing attention in the field of entrepreneurship. This approach starts with the idea that no matter what entrepreneurs say in their detailed business plans, they simply cannot make accurate predictions about many important issues. There are so many unknowns in the equation, and change happens so rapidly, that accurate prediction is almost impossible. Taking note of this fact, effectuation suggests that entrepreneurs should focus not on predicting the future but, rather, on creating it—on using the resources they have at their disposal to shape future outcomes—an idea consistent with the advice of Peter Drucker, a highly insightful and influential author and management theorist, who once remarked: "The best way to predict the future is to create it."

What are the resources that entrepreneurs can use to create—or at least change—the future? Basically, they include the characteristics, skills, and motivation of the entrepreneurs (or founding team), their knowledge and experience, and their social networks—who they know (see Chapter 5). In contrast to the formal planning represented in business plans, effectuation suggests that entrepreneurs should consider their resources they have or can have at their disposal and on the basis of these, develop a possible ends (e.g., goals). In essence, effectuation recommends that entrepreneurs should ask themselves the following question: "Given what I have and can control right now, what can I hope to accomplish?" This is in sharp contrast to the approach in business plans, which begins with specific goals (e.g., the development and sale of new products or services), and then focuses on concrete means for achieving these goals, including obtaining needed resources. Effectuation, in contrast, begins with existing resources and then considers what can be achieved with them—the way a skilled chef might begin by examining the ingredients she has available, and then coming up with ways to use them (see Figure 6.4)

In one sense, my construction of a home-made snow shovel illustrates this approach. Although I did have a specific goal in mind—clearing the snow on my driveway—I had no formal plans for doing so. Rather, I looked around the house and garage until, luckily for me, I found resources I already had that could be combined to reach this goal. Effectuation, of course, suggests that a central goal is not essential, but in this case, one existed—although I had no idea how to reach it. So perhaps this example is really one of a combination of effectuation (or effectual thinking) and a closely related process known as *bricolage*, which refers to creating something new and useful from whatever happens to be available at the moment (Baker and Nelson, 2005). The word derives from the French verb *bricoler*, which means to tinker of fiddle, and is now used to refer to what in English is known as "do it yourself"—the foundation for successful businesses such as Lowe's and Home Depot in the U.S.

While the central idea of effectuation is "avoid detailed planning and

Note: Effectuation theory suggests that it is often not necessary for entrepreneurs to engage in the kind of detailed planning included in business plans. Rather, they should consider what resources they have available, and then think about ways in which these can be used. This is similar to what a chef considering the ingredients she or he has at hand, and then planning menus on the basis of these available resources, would do.

Source: Photograph courtesy of Robert A. Baron.

Figure 6.4 *Effectuation: A gourmet example*

instead, formulate goals on the basis of what you have," it includes several other as well: (1) entrepreneurs should focus on affordable loss—what they can afford to lose, given their resources, rather than on potential profits; (2) they should form partnerships with others in order to increase the scope of their resources; and (3) they should view unexpected events as opportunities that open up additional possibilities for creative use of current resources. In other words, as one old saying puts it, "If you have lemons, don't wish for oranges—make lemonade."

As you can see, effectuation, as an approach, is almost directly opposite to the careful, detailed planning that is the core of business plans. Instead of choosing specific goals and then attempting to acquire the resources needed to reach them, entrepreneurs can, instead, imagine many different uses of the resources they possess, and take action to pursue these, remaining flexible, along the way. Do entrepreneurs ever operate in this manner? Do they ever start without a formal plan and simply try to find some way in which to use their existing skills, knowledge, experience, and so on? A growing body of evidence indicates that they do (e.g., Dew et al., 2009), and that over time, and as they gain experience, entrepreneurs tend to shift toward an effectuation approach and away from one that involves detailed planning. Apparently, as they gain experience with the complex task of making possibilities real, successful entrepreneurs come to realize that plans, in a sense, are made to be changed—so why invest great effort in making them? One problem, though, is that while the approach recommended by effectuation may well succeed in some cases, it may not be suitable in others. For instance, it would generally be very difficult to obtain financial and human resources without a detailed plan—described in a formal business plan or similar document (Baron, 2012). So, effectuation may not be the best way to proceed in many instances. Overall, then, perhaps the reasonable answer lies in a combination of effectuation, planning, and a healthy dose of willingness to improvise. In any case, it seems clear that the central idea of effectuation (Sarasvathy, 2001, 2008), that often, the future cannot be accurately predicted, but can be influenced or controlled—is certainly a very original perspective, and one well worthy of careful attention from scholars and entrepreneurs alike.

Summary of key points

Self-efficacy, the belief that we can accomplish whatever we set out to accomplish, is generally beneficial, and can enhance performance in many contexts. This is true for entrepreneurs, too: the higher they are in self-efficacy, the more likely they are to succeed in this role. However, excessive self-efficacy, especially when combined with high levels of optimism, can lead individuals to undertake tasks they are not qualified to perform, and so can actually interfere with excellent performance and success. Are entrepreneurs high risk-takers? Contrary to popular belief, evidence on this issue is mixed. One possibility is that early in the process, when risk is necessary, entrepreneurs are more willing to "take a chance" than other persons. Later, when conserving limited resources is crucial, they may be more focused on managing or reducing risk than on accepting it. Entrepreneurs may also tend to perceive less risk in a given situation than other persons.

Several aspects of personality—which refers to tendencies or preferences individuals show when they are free to act as they wish—are related to both becoming an entrepreneur and to success in this activity. These include conscientiousness, extraversion, and openness to experience. Emotions and moods (together known as affect) influence both cognition and behavior, and have important implications for entrepreneurs. In general, the tendency to experience positive affect (positive moods or feelings) often and in many situations has beneficial effects on performance, creativity, and even personal health. However, at very high levels, its effects can be detrimental (e.g., a tendency to ignore negative information that might reduce such pleasant feelings). Entrepreneurs tend to be very high in positive affect, so it is important for them to minimize the potentially detrimental effects of being "up". *Entrepreneurial passion* refers to strong positive feelings toward, and a powerful inclination to participate in entrepreneurial activities. Growing evidence indicates that such passion can have beneficial effects on entrepreneurs' behavior and may contribute to the success of their new ventures.

There is major emphasis, in most academic programs in entrepreneurship, on learning how to prepare detailed, formal business plans. Such plans are helpful in important ways, and can be an "entry card" for gaining financial resources. However, evidence on whether they are positively related to new venture success is mixed. On the other hand, current knowledge suggests that being willing and able to improvise—change strategies, plans, and actions as the need arises, is very beneficial for entrepreneurs. *Effectuation*—the view that entrepreneurs should proceed not by careful, detailed planning, but by assessing their own resources and then considering what they might accomplish with these resources—also emphasizes the importance of being flexible, and in a sense, focuses on changing the future rather than simply trying to predict it.

References

Ariely, D. (2009). *Predictably Irrational*. New York: Harper Collins.

Ashby, F.G., Isen, A.M., and Turken, A.U. (1999). A neuropsychological theory of positive affect and its influence on cognition. *Psychological Review*, 106, 529–550.

Baas, M., De Dreu, C.K.W., and Nijstad, B.A. (2008). A meta-analysis of 25 years of mood-creativity research: Hedonic tone, activation, or regulatory focus? *Psychological Bulletin*, 134, 779–806.

Baker, T., Reed, E., and Nelson, R.A. (2005). *Creating Something from Nothing: Resource Construction through Entrepreneurial Bricolage. Administrative Science Quarterly*, 50, 329–366.

Bandura, A. (1997). *Self-efficacy: The exercise of control*. New York: W.H. Free.

Bandura, A. (2012). On the functional properties of perceived self-efficacy revisited. *Journal of Management*, 38, 9–44.

Baron, R.A. (2008). The role of affect in the entrepreneurial process. *Academy of Management Review*, 33, 328–340.

Baron, R.A. (2012). *Entrepreneurship: An evidence-based view*. Cheltenham, UK and Northampton, MA, USA: Edward Elgar Publishing.

Baron, R.A., and Branscombe, N.R. (2012). *Social Psychology*, 13th edn. Boston: Allyn & Bacon (Pearson).

Baron, R.A., Hmieleski, K.M., and Henry, R.A. (in press, 2012). Entrepreneurs' dispositional positive affect: The potential benefits—and potential costs—of being "up". *Journal of Business Venturing*, 27, 310–324.

Baron, R.A., Tang, J., and Hmieleski, K.M. (2011). Entrepreneurs' dispositional positive affect and firm performance: When there can be "too much of a good thing". *Strategic Entrepreneurship Journal*, 5, 101–119.

Baron, R.A., Zhao, H., and Miao, Q. (under review, 2012). Personal motives, moral disengagement, and unethical decisions by entrepreneurs: Potential dangers of the desire for financial success.

Barrick, M.R., and Mount, M.K. (1991). The Big Five personality dimensions and job performance: A meta-analysis. *Personnel Psychology*, 44, 1–26.

Barsade, S.G. (2002). The ripple effect: Emotional contagion and its influence on group behavior. *Administrative Science Quarterly*, 41, 644–675.

Barsade, S.G, and Gibson, D.E. (2011). Group affect: its influence on individual and group outcomes. *Current Directions in Psychological Science*, 21, 119–123.

Bledow, R., Schmitt, A., Frese, M., and Kuhnel, J. (2011). The affective shift model of work engagement. *Journal of Applied Psychology*, 96, 1245–1257.

Brinckmann, J., Grichnik, D. and Kapsa, D. (2010). Should entrepreneurs plan or just storm the castle? A meta-analysis on contextual factors impacting business planning-performance relationship in small firms. *Journal of Business Venturing*, 25(1), 24–40.

Busenitz, L.W., and Barney, J.B. (1997). Differences between entrepreneurs and managers in large organizations: Biases and heuristics in strategic decision-making. *Journal of Business Venturing*, 12, 9–30.

Cardon, M.S., Gregoire, D.A., Stevens, C.S, and Patael, P. (in press). Measuring entrepreneurial passion: Conceptual foundations and scale validation. *Journal of Business Venturing*.

Cardon, M.S., Wincent, J., Singh, J., and Drnvosek, M. (2009). The nature and experience of entrepreneurial passion. *Academy of Management Review*, 34, 511–532.

Chen, X.P., Yao, X., and Kotha, S. (2009). Entrepreneur passion and preparedness in business plan presentations: A persuasion analysis of venture capitalists' funding decisions. *Academy of Management Journal*, 52, 199–214.

Ciavarella, M.A., Bucholtz, A.K., Riordan, C.M., Gatewood, R.D., and

Stokes, G.S. (2004). The Big Five and venture success: Is there a linkage? *Journal of Business Venturing*, 19, 465–483.

Dew, N., Read, S., Sarasvathy, S.D., and Wiltibank R. (2009). Effectual versus predictive logics in entrepreneurial decision-making: Differences between experts and novices. *Journal of Business Venturing*, 24, 287–309.

De Young, C.G. (2010). Impulsivity as a personality trait. In Vohs, K.D., and Baumeister, R.F. (eds), *Handbook of Self-Regulation: Research, Theory, and Applications*, 2nd edn, pp. 485–502. New York: Guilford Press.

Figner, B., and Weber, E.U. (2011). Who takes risks, when and why? Determinants of risk taking. *Current Directions in Psychological Science*, 20, 211–217.

Figner, B., Mackinlay, R.J., Wilkening, F., and Weber, E.U. (2009). Affective and deliberative processes in risky choice: Age difference in risk taking in the Columbia Card Task. *Journal of Experimental Psychology: Learning, Memory, and Cognition*, 31, 709–730.

Forgas, J.P., and George, J.M. (2001). Affective influences on judgments and behavior in organizations: An information processing perspective. *Organizational Behavior and Human Decision Processes*, 86(1), 3–34.

Fredrickson, B.L. (2001). The role of positive emotions in positive psychology: The broaden-and-build theory of positive emotions. *American Psychologist*, 56, 218–226.

Fredrickson, B.L., and Losada, M.F. (2005). Positive affect and complex dynamics of human flourishing. *American Psychologist*, 60, 678–686.

George, J.M. (1990). Personality, affect, and behavior in groups. *Journal of Applied Psychology*, 75, 107–116.

Grant, A., and Schwartz, B. (2011). Too much of a good thing: The challenge and opportunity of the inverted U. *Perspectives on Psychological Science*, 6, 61–76.

Hmieleski, K.M., and Baron, R.A. (2009). Entrepreneurs' optimism and new venture performance: A social cognitive perspective. *Academy of Management Journal*, 52, 473–488.

Hmieleski, K.M., and Corbett, A.C. (2008). The contrasting interaction effects of improvisational behavior with entrepreneurial self-efficacy on new venture performance and entrepreneur work satisfaction. *Journal of Business Venturing*, 23, 482–496.

Hmieleski, K.M., Corbett, A.C., and Baron, R.A. (in press, 2012). The relationship between entrepreneurs' improvisational behavior and firm performance: An interactional study of dispositional and environmental moderators. *Strategic Entrepreneurship Journal*.

Isen, A.M. (2000). Positive affect and decision making. In Lewis, M. and Haviland-Jones, J.M. (eds), *Handbook of Emotions*, 2nd edn, pp. 417–435. New York: Guilford Press.

Judge, T.A., and Ilies, R. (2004). Is positiveness in organizations always desirable? *Academy of Management Executive*, 18(4), 151–155.

Kaplan, S., Bradley, J.C., Luchman, J.N., and Haynes, D. (2009). On the role of positive and negative affectivity in job performance: a meta-analytic investigation. *Journal of Applied Psychology*, 94, 162–176.

Khaire, M. (2010). Young and no money? Never mind: The material impact of social resources on new venture growth. *Organization Science*, 21 (January–February), 168–185.

Knight, A.P. (2011). Mood at the midpoint: How team positive mood shapes team development and performance. In Toombs, L.A. (ed.), Proceedings of the 71st Annual Meeting of the Academy of Management, San Antonio, Texas.

Lange, J.K.E., Mollow, A., Pearlmutter, M., Singh, S., and Bygrave, W.D. (2007). Pre-start-up formal business plans and post-start-up performance: A study of 116 new ventures. *Venture Capital*, 9, 237–256.

Larson, R., and Buss, D.H. (2009). *Personality Psychology*. New York: McGraw Hill.

Lyubomirsky, S., King, L., and Diener, E. (2005). Benefits of frequent positive affect. *Psychological Bulletin*, 131, 803–855.

Markman, G.D., Balkin, D.B., and Baron, R.A. (2002). Inventors and new venture formation: The effects of general self-efficacy and regretful thinking. *Entrepreneurship Theory & Practice*, Winter, 149–165.

Melton, R.J. (1995). The role of positive affect in syllogism performance. *Personality and Social Psychology Bulletin*, 21, 788–794.

Miner, J.V., and Raju, N.S. (2004). When science divests itself of its conservative stance: The case of risk propensity differences between entrepreneurs and managers. *Journal of Applied Psychology*, 89, 3–13.

Mount, M.K., Barrick, M.R., and Strauss, J.P. (1994). Validity of observer ratings of the big five personality factors. *Journal of Applied Psychology*, 79, 272–280.

Mullins, J., and Komisar, R. (2009). *Getting to Plan B: Breaking through to a better business plan*. Cambridge, MA: Harvard University Press.

Murnieks, C.Y., Mosakowski, E., and Cardon, M.S. (in press, 2012). Pathways of passion: Identity centrality, passion, and behavior among entrepreneurs. *Journal of Management*.

Oishi, S., Diener, E., Choi, D.W., Kim-Prieto, C., and Choi, I. (2007). The dynamics of daily events and well-being across cultures: When less is more. *Journal of Personality and Social Psychology*, 93: 685–698.

Sarasvathy, S, (2001). Causation and effectuation: Toward a shift from economic inevitability to entrepreneurial contingency. *Academy of Management Review*, 26, 243.

Sarasvathy, S.D. (2008). *Effectuation: Elements of Entrepreneurial Expertise*. Cheltenham, UK and Northampton, MA, USA: Edward Elgar, New Horizons in Entrepreneurship Series.

Simon, M., Houghton, S.M., and Aquino, K. (2000). Cognitive biases, risk perception, and venture formation: How individuals decide to start companies. *Journal of Business Venturing*, 15, 113–134.

Stewart. W.H., and Roth, P.L. (2001). Risk propensity differences between entrepreneurs and managers: A meta-analytic review. *Journal of Applied Psychology*, 86, 145–153.

Stewart, W.H., and Roth, P.L.L. (2007). A meta-analysis of motivation differences between entrepreneurs and managers. *Journal of Small Business Management*, 45, 401–421.

Vallerand, R.J., Blanchard, C., Mageau, G.A., Koestner, R., Ratelle, C., Leonard, M., Gagne, M., and Marsolais, J. (2003). Les passions de l'ame: On obsessive and harmonious passion. *Journal of Personality and Social Psychology*, 85, 756–767.

Weiss, H.M., and Cropanzano, R. (1996). Affective Events Theory: A theoretical discussion of the structure, causes and consequences of affective experiences at work. *Research in Organizational Behavior*, 18, 1–74.

Zhou, J., and George, J.M. (2007). Dual tuning in a supportive context: joint contributions of positive mood, negative mood, and supervisory behaviors to employee creativity. *Academy of Management Journal*, 50, 605–622.

Zhao, H., and Seibert, S.E. (2006). The Big Five personality dimensions and entrepreneurial status: A meta-analytical review. *Journal of Applied Psychology*, 91, 259–271.

Zhao, H., Seibert, S.E., and Hills, G.E. (2005). The mediating role of self-efficacy in the development of entrepreneurial intentions. *Journal of Applied Psychology*, 90, 1265–1272.

Zhao, H., Seibert, S.E., and Lumpkin, G.T. (2010). The relationship of personality to entrepreneurial intentions and performance: A meta-analytic review. *Journal of Management*, 36, 381–404.

7 Making effective decisions—and avoiding cognitive traps

Chapter outline

The basic nature of decision-making—and why it is far from totally rational
Image theory: an intuitive perspective on decision-making
Cognitive traps decision-makers should (must!) avoid
Implicit favorites: the powerful—and lasting impact—of initial preferences
The person sensitivity error: the self-serving bias revisited
Failing to recognize when it is time to get out: the dangers of escalating commitments
Decision-making strategies—and why some are better than others
Maximizing versus satisficing: the potential downside of seeking perfection
Signal detection theory: weighting the relative value of potential outcomes
Group decision-making: benefits and costs
Group polarization: why groups sometimes go "off the deep end"
The failure to share information known only to some members
Biased processing of information: do group members seek accuracy—or being "right"?

* * *

Life is the sum of all your choices.
(Albert Camus)

When making a decision of minor importance, I, always . . . consider all the pros and cons. In vital matters, however . . . the decision should come from somewhere within ourselves, the deep inner needs of our nature.
(Sigmund Freud)

Whenever you see a successful business, someone once made a courageous decision.
Peter Drucker

Making decisions is, if nothing else, hard work (Hardman, 2009); the more complex the decisions, the more important, the consequences, and the more quickly we have to make them, the more challenging making them becomes. Yet, as Camus suggests, life—and we could add success—is, in a sense, the sum of all our decisions. Entrepreneurs, of course, are no exception to this general rule. They must make countless decisions, often in the face of incomplete knowledge or information and often under great time pressure. Further, they must do so under conditions where there are no simple or ready-made solutions (after all, they are breaking new ground!), and where information necessary for assessing the risks involved is either lacking, or perhaps even unattainable (e.g., how can one know with certainty that potential customers will actually like, and buy, a new product or service?). In a sense, then, decision-making takes on extra importance, and is even more challenging, for entrepreneurs than for persons in many other fields or occupations.

This reasoning suggests that entrepreneurs should seek to add the ability to make decisions effectively and efficiently to their list of essential skills—that they should definitely include it among the tools that will help them change the *possible* into beneficial reality. In this chapter, we will review information relevant to this important task. Specifically, we will proceed as follows. First, we will examine the basic nature of decision-making, and briefly review two contrasting models of how this process unfolds. Next, we will turn to important pitfalls individuals face when making decisions—factors that can get in the way of achieving the goal of effective decisions made in an efficient manner. Third, we will examine what might be termed individual "styles" or approaches to making decisions, and explain why, on the basis of existing evidence, some are more effective than others. Finally, we will focus on decision-making in groups—a situation often faced by founding teams of entrepreneurs. Here, we will see that groups, too, are subject to important forms of error—ones stemming from processes that can lead them, if not recognized and countered, to truly disastrous decisions.

The basic nature of decision-making: and why it is far from totally rational

How are decisions actually made? If you think for a moment, you may realize that this depends strongly on how important they are. If a decision is fairly trivial, such as "Should I have mushrooms or sausage on my pizza—or both?" it is usually made quickly and relatively effortlessly. But as the importance of the decision increases, so, too, does our willingness to invest lots of cognitive effort in making it; and this makes very good sense: important decisions are ones that have significant outcomes, so we want them to be good ones. This important dimension is largely ignored, however, by one influential model of decision-making—the

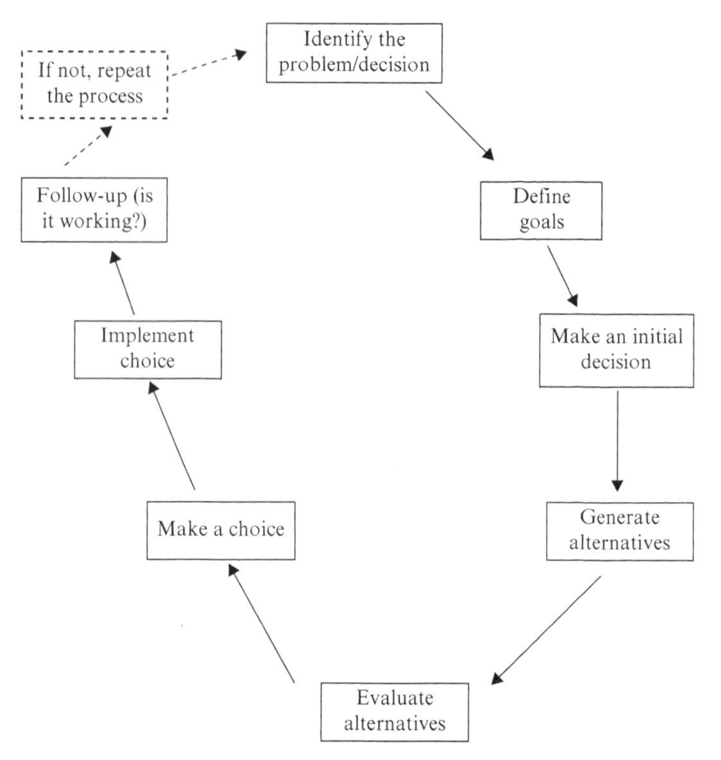

Note: In theory, decision-making proceeds as shown here—through a very rational and analytical process. In fact, many cognitive tendencies, errors, and biases make it difficult—if not impossible—to be entirely rational where decision-making is concerned.

Source: Robert A. Baron.

Figure 7.1 *The rational analytic model*

rational analytic model. Basically, this model (which derives primarily from economic theory as it existed in the past), assumes that decision-making is a very rational process in which individuals engage in three distinct but related steps. First, they formulate the decision—identify the problem, define their goals, and decide how to go about making the decision—what procedures to use. Next, they actively generate alternatives, evaluate them, and make an initial choice. Third, they implement the decision—they put it into action, and then, finally, follow-up by monitoring the effectiveness or success of the decision. These basic steps are illustrated in Figure 7.1.

Do individuals actually make decisions in this rational and analytic manner? Sometimes, but certainly not always. When the decisions are trivial, they do not follow this careful, logical formula: the cognitive effort is not justified. When decisions are of moderate to high importance—for instance, which of two similar jobs to accept—they may well do so. But as noted by Freud in the quotation above, when faced with truly major, life-shaping decisions, many people seem to operate in a more intuitive manner. For instance, in deciding to make a life-time commitment to another person through marriage, few individuals proceed by means of the steps above—by considering all the alternatives, the pros and cons of each potential partner, and so on. Rather, they tend to go with their "heart"—what their feelings and emotions tell them. Similarly, when considering whether to launch a new venture or not, it may be difficult, if not impossible, for entrepreneurs to follow a purely analytic, rational process. Once again, as we noted in Chapter 6, emotions play a key role in our cognitive processes, and this is certainly true for decision-making. That is a crucial point, and one to which we will return at several points in this chapter.

Image theory: an intuitive perspective on decision-making

The powerful impact of our emotions is only one reason why decision-making is not always an analytical and rational process. As we saw in Chapter 3, our thinking—and that includes decision-making—is subject to many forms of error and bias. We will consider some of these, including ones we have not discussed previously, in later sections. Here, however, we wish to briefly mention another, and sharply different, view of decision-making—one that takes account of the fact that many decisions are made in a more rapid and intuitive manner than the rational analytic model describes. This view is known as *image theory* (e.g., Beach and Mitchell, 1990; Dunegan, 1995), and in essence, it suggests that when individuals are faced with a decision—especially one about choosing a particular course of action (e.g., should they seek a patent for an idea?) they first compare this course of action with mental images of its potential outcomes. Will it help them attain the goals they seek? Is it compatible with their personal values? Then, they compare potential courses of action to see which provides the best match—which is more likely to help them achieve their goals, and so on. Both steps occur rapidly and in an intuitive manner: detailed analysis does not take place. If you have ever been in a situation where you made a decision to pursue some course of action and because you thought something like "This just feels right", you have first-hand experience with image theory. And if you have ever rejected a course of action that offered many obvious advantages and then thought "I don't know . . . it just didn't seem to be right thing to do" (e.g., rejected what seemed to be a good job offer because something about it just did not seem appealing), you have had additional experience with this approach to making decisions.

Do people actually operate in this manner? Research findings indicate that they do (Dunegan, 1995). Whether the steps above are the precise ones individuals following making decisions, however, this perspective is valuable if only because it calls attention to the fact that many decisions are not made in a purely logical manner. On the contrary, they reflect intuition—feeling, reactions and thoughts we cannot readily put into words—as strongly, if not more strongly, than the factors contained on any list of the "pluses" and "minuses" of various options or choices.

Cognitive traps decision-makers should (must!) avoid

In a sense, a sub-theme of this entire book—one we have emphasized at several points—is this: "Our cognitive skills and capacities are nothing short of amazing—but they are also far from perfect." Taking account of this important fact, we have called attention to the many ways in which the limitations of our memories, our information processing capacities, and even our own creativity, can often lead us astray, and generate many unexpected effects. Decision-making is a cognitive process, so it, too, is subject to these problems. Since, as Albert Camus suggests, our lives are the sum of our choices or decisions, we will now echo this "cognitive limitations" theme once more, by calling attention to several sources of bias and error that appear to be built into our cognitive systems, and often—if not held in check—can result in truly devastating decisions and outcomes.

Implicit favorites: the powerful—and lasting impact—of initial preferences

As we noted earlier, decision-making is often hard work. When the decisions are important, and have potentially life-changing and long-lasting implications, people work hard to be as rational as they can. But how rational is that? Can they really list all the alternatives and evaluate them fairly and adequately? From the discussion so far, you can probably guess that the answer is "Not always". Several factors make this process easier to imagine than to achieve. One of these involves the powerful impact of initial decisions.

When individuals face an important decision and consider all the obvious alternatives, they often find that they have an initial preference. This is known as the *implicit favorite*, because often, people do not admit, even to themselves, that they are leaning in a particular direction. They want to at least believe that they are "keeping their options open". But in fact, research on decision-making indicates that this more of an illusion than a reality. In fact, it appears, most people ultimately tend to go with their initial inclination or preference. Why? Because in the light of this initial preference, they often tend to direct reduced attention to the other alternatives, think about them less often, but actively seek reasons why their initial preference—the implicit favorite—is, in fact, the best one. In a sense, their initial preference triggers the *confirmation bias*—the tendency to confirm our own beliefs or views by focusing primarily on information that supports them; and in decision-making situations, one good way of doing this is to strengthen support for the initial favorite, while downplaying that for the other choices. Many studies provide evidence for this process, and in fact, when we turn to decision-making by groups, we will encounter it again in the form of a strong tendency for decision-making groups to choose the option initially favored by most members—even in the face of growing evidence that it is not the best choice. The moral? Watch out for initial preferences because they often serve to "lock out" full consideration of other, and perhaps better, options.

The person sensitivity error: the self-serving bias revisited

Political wisdom suggests that at least in the U.S., the fate of incumbent Presidents, and that of candidates for this high office, are strongly influenced by the state of the economy. In recent events, President Bush (the senior) lost the 1992 election to Bill Clinton, in part because the U.S. economy was sinking rapidly during his last year in office. In the 2012 election, the U.S. economy was once again very weak, which led many political pundits to predict (wrongly!) that President Barack Obama's chances of winning re-election were much reduced. In fact, Presidents have very limited powers to influence the economy; and even if the policies they pursue do have such effects, these are not generally visible for years. Yet, the public seems to attribute bad economic times to whoever is currently in office. In a sense, this is an offshoot of the self-serving bias, described in Chapter 3. That term refers to a powerful and general tendency to attribute positive outcomes to internal causes (e.g., our own effort, talent, or skill), but negative outcomes to external causes (e.g., other persons, forces beyond our control, bad luck; Greenberg and Baron, 2009), In the context of decision-making, this tendency has been described as the *person sensitivity error*—a powerful tendency to assign too much blame to others when things (e.g., the economy) go badly, but too much credit to them when things are going well. This, in turn, can strongly influence decisions—political and otherwise.

For instance, suppose an entrepreneur has obtained financial backing from venture capitalists (VC). This financial support is usually provided over several stages, as the new venture achieves various goals (e.g., initial sale of its products or services, certain levels of cash flow, etc.). Imagine, now, that the company fails to meet one of these goals. Who or what is to blame? Because of the powerful impact of the person sensitivity error, the VC may tend to accentuate the role of the entrepreneur or founding team, and decide not to provide additional funding. This might well be true even if the factors generating the company's problems have little or nothing to do with the entrepreneurs and are totally beyond their control. In short, even skilled decision-makers such as highly experienced VC may be subject to this bias, and so make decisions that are inappropriate in the light of actual circumstances.

Failing to recognize when it is time to get out: the dangers of escalating commitments

In our previous discussion of important cognitive errors (Chapter 3), we described the powerful effect of *sunk costs*—the tendency to stick with bad decisions once they are made. Another term for this effect is *escalation of commitment*, and as it suggests, one reason people get trapped in bad decisions is that once these choices are made and resources (e.g., time, energy, effort) are invested in trying to implement them, individuals soon reach the point where, cognitively, they feel that they have "too much invested to quit". In other words, they conclude that they cannot readily reverse a decision because they are now too deeply committed to it. No one wants to admit that they were wrong, so they avoid doing so and forge ahead, hoping that the situation will improve.

This is a major danger for decision-makers, and relates also to one aspect of metacognition (our understanding of our own thinking): the ability to know when it is time to "get out"—to withdraw from a failing course of action or strategy. This is one major reason why so many new ventures fail: the entrepreneurs believe—often passionately—in their ideas, and in the business models they have formulated to develop them. As a result, they tend to be overoptimistic (another important cognitive bias), and ignore or downplay evidence that things are not working our as they planned; rather, they continue to hope that the situation will soon (perhaps miraculously?) improve.

Research findings indicate that knowing how much time and effort to spend on a chosen course of action before admitting that it was the wrong choice, is major predictor of success in many situations (e.g., Jarvstad et al., 2012). For instance, assume an entrepreneur is developing a new product she plans to bring to market, but things are not going well—problems remain, despite the investment of time, effort, and sizeable financial funds, and these problems are not decreasing. In such a situation it is all too easy to fall victim to sunk costs: why quit now, the entrepreneur may reason, when so much has already been invested in developing this product and success may lie just around the corner? Being able to make such choices—in essence, being able to accurately judge when it is indeed time to "get out"—is crucial; in fact it can make the difference between success and failure. Fortunately, we seem to be quite good at this task (Jarvstad et al., 2012) in relatively simple situations. In more complex ones, however, we

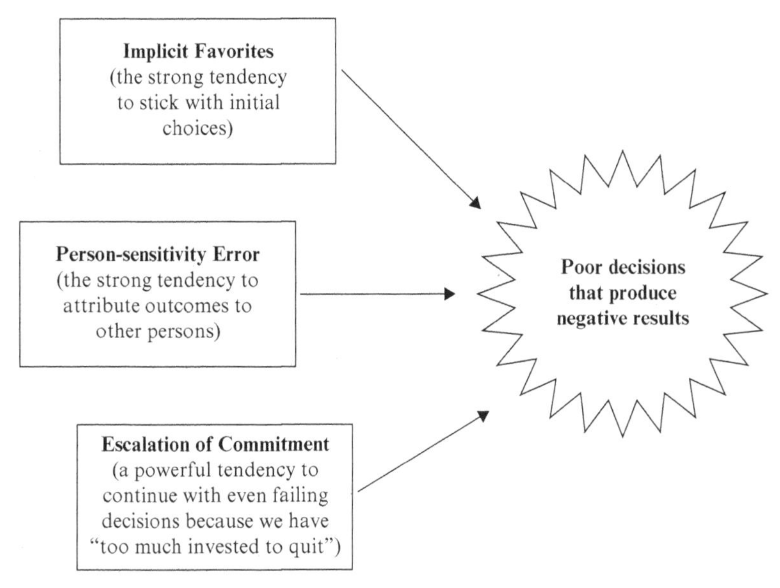

Note: Several different processes can interfere with decision-making by individuals, and result in choice with devastating results. The ones shown here are among the most important.

Source: Robert A. Baron.

Figure 7.2 *Cognitive errors that can lead decision-making badly astray*

are more susceptible to major errors (Kahneman, 2003). In view of such findings, it seems clear that effective decision-making involves not simply choosing what seems to be the best available option, but also being able to recognize when this initial choice was not the optimal one, and having the courage—and flexibility—to admit this fact, and change course as quickly and effectively as possible (see Figure 7.2 for a summary of important cognitive errors that can lead decision-makers into serious, and sometimes devastating, errors).

Decision-making strategies—and why some are better than others

In any given situation, there are many strategies for reaching a decision. As described earlier, one approach—the rational analytic strategy—is to identify all possible choices, evaluate them systematically, and proceed from there to making a decision. Another possibility, explained earlier, is to operate in a more intuitive manner, quickly comparing possible alternatives with important goals and estimating the likelihood that each alternative will enhance progress toward these goals. These basic models—informative as they are—are only part of the total picture in terms of how individuals actually make decisions. Two other perspectives help provide further details on how this crucial process actually unfolds.

Maximizing versus satisficing: the potential downside of seeking perfection

When you have to make an important decision, do you search carefully through all possible alternatives looking for the best one—the perfect choice? Or do you, instead, search through alternatives until find the first one that "works"—the first one that you find acceptable, even though you recognize that it is probably not ideal? Research evidence indicates that in making decisions, individuals tend to adopt one or the other of these contrasting strategies. The first is known as *maximizing*, since it is focused on finding the very best alternative. People who typically adopt this approach are described as being *maximizers*. The second approach is known as *satisficing*, because it involves searching for something that will work, even though it is not perfect or ideal (Schwartz et al., 2002; Schwartz, 2004). People who prefer this latter strategy are known as *satisficers*.

The really interesting thing about these two strategies for making decisions is that they offer a mixed pattern of costs and benefits. Maximizers engage in very careful and diligent comparison of all existing alternatives; this means that all else being equal, they tend to make good decisions—perhaps better ones than satisficers. In other words, their search for perfection is, in one sense, rewarded. But since maximizers seek perfection (or at least, the very best possible choices), they tend to be dissatisfied with their decision after they make them. Are these choices really the best? It is impossible to know for certain. Satisficers, in contrast, are generally pleased with their decisions since they have saved a great deal of effort by choosing the first one that is "good enough".

The overall effect is that satisficers tend to be happier not only with their decisions, but with their lives generally. They tend to be higher in what psychologists term *subjective well-being* than maximizers (see Chapter 8). But what does all this tell us about entrepreneurs—individuals who are attempting to develop something new and better? In general, they face situations that are not only unpredictable, but also ones that offer little opportunity to engage in the kind of detailed analysis required for maximizing. Opportunities often exist for only a limited amount of time before someone recognizes and develops them. So overall, a satisficing strategy may be better for entrepreneurs. Of course, there is a downside to this approach: the decisions made may be far from ideal and require later adjustments. If satisficing is combined with the willingness to improvise (described in Chapter 6), however, the result may be the best combination of speed and effectiveness available in the difficult and sometimes chaotic

Note: Existing evidence indicates that although people generally like to have choices, too many alternatives can generate negative feelings stemming from opportunity costs (recognizing that by making one choice we have given up the benefits of others), and unmet expectations. Unfortunately, as shown here, life in many societies offers what may well be too many choices!

Sources: Fotolia 11919665; 20952811.

Figure 7.3 *Is "more always better" where number of choices is concerned?*

world of entrepreneurs—a world the entrepreneurs are attempting to "shake up" in some manner.

Is having many options always desirable?

Before turning to another basic approach to decision-making, it is both interesting and important to pause briefly and consider a question closely related to maximizing and satisficing: can individuals have too many choices? In general, as human beings, we tend to find choice desirable—recent evidence indicates that anticipating having choice, as well as actually having it, stimulates brain regions involved in feelings of pleasure and reward (Leotti and Delgado, 2012). But, to echo a theme mentioned at several points in this book, can there be "too much of a good thing" where choice is concerned? Growing evidence indicates that this may be true. As noted by Schwartz (2004) and others, people in developed nations such as the U.S. currently have a huge number of choices in many situations, especially ones involving the purchase of various products. Walking down the aisles of a modern supermarket or visiting a large shopping mall makes this point crystal clear. By actual count, consumers can choose between 273 brands and types of breakfast cereal and even with respect to larger purchases such as automobiles, the number of available models has increased greatly in recent decades (see Figure 7.3). Does this abundance of choice enrich our lives and make us happier? Growing evidence (Schwartz, 2004) indicates that it does not. In fact, there are several forces that, operating together, can lead to negative rather than positive emotions and feelings in the face of a large number of choices. First, of course, to the extent individuals are maximizers, they may face total decision paralysis: how can they evaluate all the possible choices and choose the best one? And for both maximizers and satisficers, other processes begin to operate. As the number of choices rises, so, too, do *opportunity costs*. These refer to the fact that when we choose one option, we must usually forego the others. Unfortunately, we are often all too aware of the potential benefits we have lost by choosing one job, one automobile, one home or apartment over others. The greater the number of choices, the greater the likelihood that such effects will occur.

Similarly, when individuals have many choices, their expectations about the outcomes they will gain are very high. After all, with so many options, how can they go wrong? In fact, though, the higher our expectations, the more likely we are to be disappointed—to find the choice we have made to be less desirable or effective than we anticipated. Finally, and related to these effects, we may experience more regret and be more likely to imagine what might have happened if we had made other choices—accepted a different job, pursued a different career, married a different person, had more (or fewer children), and so on. The result of these effects is that the relationship between number of choices available and personal happiness may be curvilinear: up to a point, more choice generates positive emotions, but beyond some point, more choice results in less positive, and more negative, feelings. So perhaps there is a paradox in choice: more is not always better and in fact, too many choices may leave the persons facing them—even satisficers—with the feeling that no matter what decisions they have made, they could well have done even better!

Signal detection theory: weighting the relative value of potential outcomes

As we noted in Chapter 3, a key task for entrepreneurs is identifying opportunities—perceived means of generating value (i.e., profit or other benefits) that are not currently being exploited and are viewed, in a given society, as desirable or, at least, socially acceptable. Opportunity recognition, in turn, involves several basic cognitive processes, such as *pattern recognition*—connecting the dots between various events and trends to form a recognizable pattern that then provides the basis for a business opportunity (Baron, 2006). But how do entrepreneurs decide whether the opportunities they have observed are actually real ones? Opportunities are not always clear or obvious. Does the artificial leaf developed by Professor Nocera (see Chapter 2), really constitute an opportunity? What about the hiccup pop described in Chapter 3? Often, it is impossible to tell in

Stimulus is Present	Stimulus is Absent
Hit Correct identification of a stimulus (e.g., opportunity is present and is recognized)	**False Alarm** Opportunity is not present, but is interpreted as being present
Miss Opportunity is present, but is not recognized as being present	**Correct Rejection** Opportunity is not present and is recognized as being absent

Note: Signal detection theory suggests that many decisions—for example, whether a recognized business opportunity is real and therefore worth pursuing—or illusory, and therefore a potential blind alley—involve the four possibilities shown here.

Source: Robert A. Baron.

Figure 7.4 *Signal detection theory: an important perspective on decision-making*

advance. There is an important decision-making perspective, however, that seems to shed important light on such decisions.

This perspective is known as *signal-detection theory* (Swets, 1992), and it addresses a basic and important question: how do individuals decide that a stimulus is actually present in a given situation? This sounds abstract but in fact, it is basically the task entrepreneurs face when trying to decide whether to pursue a business opportunity they have identified. Is it really there? Does it really exist? In such situations, four possible outcomes exist: the stimulus (opportunity) is actually present and is recognized accurately (in signal theory, that is known as a hit or correct identification); the opportunity is present but is not recognized (a miss); the opportunity is absent but the entrepreneurs concludes, erroneously, that it is present (a false alarm); or the opportunity is absent and the entrepreneur correctly concludes that it is absent (a correct negative or correct rejection). The goal, of course, is to maximize decision accuracy by pursuing only hits—opportunities that are actually present, that do exist (see Figure 7.4).

The theory notes that many factors influence the relative rate at which individuals experience hits, misses, and false alarms in any given situation, but among these, the most relevant to the present discussion is the subjective criterion individuals apply to the task. The higher this criterion the less likely is the entrepreneur to conclude that the opportunity is real and worth pursuing; the lower it is, the more likely she or he is to decide that the opportunity is real, and take action to develop it. Consider the situation faced by an entrepreneur who believes that she has identified an opportunity for a profitable new venture. The venture is one that she can start in her spare time and for which little or no capital is needed. As a result, she may set her subjective criterion for concluding "This is a good business opportunity" quite low: the costs of a false alarm are minimal (a little wasted time and effort) relative to the potential gains of a hit.

In contrast, consider another entrepreneur who has recognized an opportunity that cannot be pursued on a part-time basis and for which large amounts of start-up capital are required. Under these circumstances, the entrepreneur will probably set her criterion for concluding "This is a real or good opportunity" somewhat higher: the costs of a false alarm are very high and potential rewards are reduced by the large proportion of the business that will be owned by investors. In short, potential costs and benefits of starting a new venture determine where prospective entrepreneurs set their criteria for concluding that an opportunity they have perceived is real and therefore worthy of active development.

Here is another example that may help clarify the nature of this process and how it captures important aspects of decision-making. Imagine a radiologist who, while examining MRIs or CAT scans, searching for a possible tumor in a patient finds that even with highly sophisticated equipment, the scans are inconclusive—it is not clear whether a tumor exists or not. What does the radiologist do? Where does he or she set the subjective criterion for concluding "Yes, there is a tumor present"? If the patient is 90 years old and unlikely to survive major surgery, the radiologist may set the criterion very high: the costs of a false alarm (performing unnecessary surgery) are very high (the patient might well die during surgery), perhaps much higher than those of a miss (not detecting a tumor that actually exists; the patient is already 90 years of age). If the patient is 25 years old and in good health, however, the radiologist may set the criterion much lower: the costs of a miss are much higher than those of a false alarm, since the patient can probably survive major surgery very well and has many years of life ahead of her. Returning once more to entrepreneurs, signal detection theory suggests that entrepreneurs who are strongly motivated to minimize risks and to avoid pursuing false alarms, may set their subjective criteria relatively high, while those who are relatively tolerant of risk and more concerned about overlooking bona fide opportunities may set their criteria somewhat lower (see, e.g., Busenitz and Barney, 1997; Krueger and Brazael, 1994; Stewart and Roth, 2001). Similarly, entrepreneurs who are high in optimism may set their subjective criteria low, with the result that they experience many false alarms.

In short, signal detection theory adds to our understanding of decision-making by calling attention to the fact that in situations where decisions are complex and difficult to make, the general context in which they occur often strongly influences the choices made.

Group decision-making: benefits and costs

Groups of individuals working together are called on to perform many tasks—everything from playing music in an orchestra through performing complex surgical operations. One of the most important activities groups perform, though, is making decisions. Governments, large corporations, military units—these and many other organizations call on groups to make important decisions. Why? Why do they not just allow the individual involved to reach their own decisions and make their own choices? One answer is that it is widely believed that groups tend to make better decisions than individuals. True, they may take longer to reach them, but the basic idea is that the decisions, once reached, will be more accurate and effective than those formulated by individuals working alone.

Is this true? If so, then entrepreneurs, too, should entrust key decisions to groups—for instance, the founding team working together, or entrepreneurs and venture capitalists, or entrepreneurs and key employees. If not, then perhaps many decisions should be entrusted to the persons most directly involved with them. Given the complexity of these questions, it is not surprising that research comparing the effectiveness of decisions by individuals and groups yields a mixed picture (Mojzisch and Schulz-Hardt, 2011). On the one hand, groups do have important potential advantages: members can pool their knowledge and skills, can act as "brakes" on each other's extreme views, and can, perhaps, make better plans for implementing the decisions once they are adopted. The key question, then, is this: are such benefits actually realized? And once again, the answer is: only sometimes. We will now consider reasons why groups do not always make better decisions than individuals. (We already considered one reason for poor decisions by groups in our earlier discussion of groupthink, see Chapter 5.)

Group polarization: why groups sometimes go "off the deep end"

One reason why many important decisions are assigned to groups is that, as noted above, it is widely assumed that they are less likely to make extreme choices than individuals. That seems logical, but in fact, decades of research on this issue indicate that groups are often more likely to "go off the deep end" than individuals. This tendency is known as *group polarization*, and it has been found to operate in a wide range of decision-making groups. In essence, it refers to the fact that whatever position is initially favored by the group (i.e., favored by a majority of its members), tends not merely to be adopted as the group's final decision, but also tends to be adopted more extremely. For instance, if a majority of the members are slightly in favor of one option, this is likely to be adopted, and in a more extreme form than was initially true. For instance, consider a team of five founding entrepreneurs discussing the question of whether to form an alliance with a larger company. Assume that initially, three of the members are in favor of this option, with one being slightly favorable and the other two being moderately so. Two other members are slightly negative. What will happen when the group discusses this decision at length? Group polarization suggests that after their discussions, the group will be moderately to strongly in favor of forming the alliance. In other words, it will have shifted toward a position stronger (more extreme in the favorable direction) than any of the members held previously. Why? There is nothing mysterious going on. This shift reflects the fact that since a majority of the members were in favor of forming an alliance, most of the arguments made during discussions also favor this choice. The result is that those who were slightly against this course of action become more positive, and those who were already favorable—hearing these arguments—move even further in this direction. The overall result is that the group's decision is more extreme than the decisions recommended initially by its individual members. In a sense, the members convince each other that the position initially favored by most members is the correct one.

The same process seems to occur in a wide range of decision-making groups: whatever their initial preference, it is strengthened after group discussion of the decision options. This effect is very powerful: in fact, history is replete with examples of decision-making groups that adopted extreme decisions with devastating effects, such as the decision by President Johnson and his "expert" advisers to escalate American involvement in Viet Nam, President Bush's decision (again, supported by most of his cabinet) to invade Iraq, the decision of the Board of Directors of Apple Computers to fire Steve Jobs in 1990, and many other decisions too numerous to mention. Since new ventures generally have severely limited resources and cannot easily recover from the consequences of bad decisions, group polarization is definitely a potential danger founding teams of entrepreneurs should consider carefully as they proceed.

The failure to share information known only to some members

One of the key advantages supposedly held by groups involves the "pooling of resources": ostensibly, group members each have unique stores of knowledge and skills and can share these during the decision-making process. Again, this sounds very reasonable and certainly occurs in some groups. However, research that has observed the actual process through which decisions are reached by groups indicates that in many instances, the members do not exchange or share unique information. Rather, they tend to focus on information known to all or most members—information that is already shared. This is not a problem if this shared information does indeed point to the best decision. But if groups spend much of its time going over (and over!) information most members already have, they may fail to share information held by only one or a few members—information that, perhaps, would point to a different and superior decision. Again, research findings offer support for this reasoning: in medical decisions made by teams of physicians and medical students, the greater the extent to which the teams discussed unshared information, the better the decisions they made (e.g., the more accurate the diagnoses; Gigone and Hastie, 1997).

How can groups avoid focusing only on information they already share? One technique involves having members prepare lists of potential choices as well as the pros and cons associated with each *prior* to the group discussion. Then, if some of these alternatives are not discussed, they can be introduced into the group conversation by individual members. Another approach is known as the devil's advocate technique. In this technique (e.g., Hirt and Markman, 1995), one member is assigned the task of disagreeing with and criticizing whatever decision the majority of members seems to favor. By doing so, the "devil's advocate" encourages members to think more carefully about the reasons why they support certain views, and this, in turn, may bring unshared information to into focus.

Biased processing of information: do group members seek accuracy—or being "right"?

Yet another reason why groups do not necessarily make better decisions than individuals involves what is known as *biased processing of information*. It is often assumed that when they discuss various decisions, group members are motivated primarily to choose the best one. That is certainly true in many instances. But as we noted in Chapter 2, complex human behavior generally reflects the simultaneous operation of several motives rather than a single motive. A quest for accuracy and the best decision may be only one of these. Another is the desire, by individual members, to be "right"—to have their positions adopted by the group. So, instead of acting like scientists on a search for truth, they behave more like attorneys, seeking support for their positions (clients). When this happens, "winning" (having one's own views accepted by the group) takes precedence over decision accuracy, with the result that the choice adopted may, in fact, not be the best one available. This tendency, too, can be counteracted by groups that are aware of, and vigilant, for its occurrence. One means of doing this is to separate members from their personal positions or preferences insofar as possible. This can sometimes be accomplished, by having written (rather than open) "straw polls" to assess the group's current preferences, and by emphasizing the fact that the decisions made have important consequences for all members—not just those whose views are the ones adopted. Again, though, reducing this potential source of bias is far from easy or automatic.

In sum, there are several reasons why groups do not necessarily make good decisions—or ones better than those that would be reached by individuals acting alone. The key task for groups of entrepreneurs, as for all decision-making groups, is to maximize the potential benefits of group process (e.g., pooling resources, achieving greater commitment to implementing the decisions reached), while holding potential sources of error at bay. That is certainly a complex task, but one well worth the effort if it leads to better (i.e., more effective) decisions. (See Figure 7.5 for an overview of the processes that interfere with effective group decisions.)

Summary of key points

One model of decision-making views it as a largely rational and analytic process, in which individuals identify and evaluate alternatives in a systematic manner. Another suggests that decision-making is more intuitive, involving comparison of potential outcomes associated with various choices with important goals. Existing evidence indicates that in fact, decision-making cannot be totally rational, because it is subject to several important cognitive errors. One of these involves implicit favorites—the fact that individuals generally choose the options they initially favor, despite evidence that it might not be the most effective. Another is the *person sensitivity bias*—a tendency to assign too much blame to others when things are going badly, but too much credit to them when things are going well. Another important source of error in decision-making is the inability to recognize when it is time to withdraw from bad decisions (escalation of commitment).

In making decisions, individuals tend to be either *maximizers*, who week the best choice, or *satisficers*, who seek the first solution that works. Satisficers are happier with the decisions they make than maximizers, and also happier with their lives in general. As human beings, we tend to find having choices desirable, but research evidence indicates that too much choice—too many options—can have negative effects on our emotions

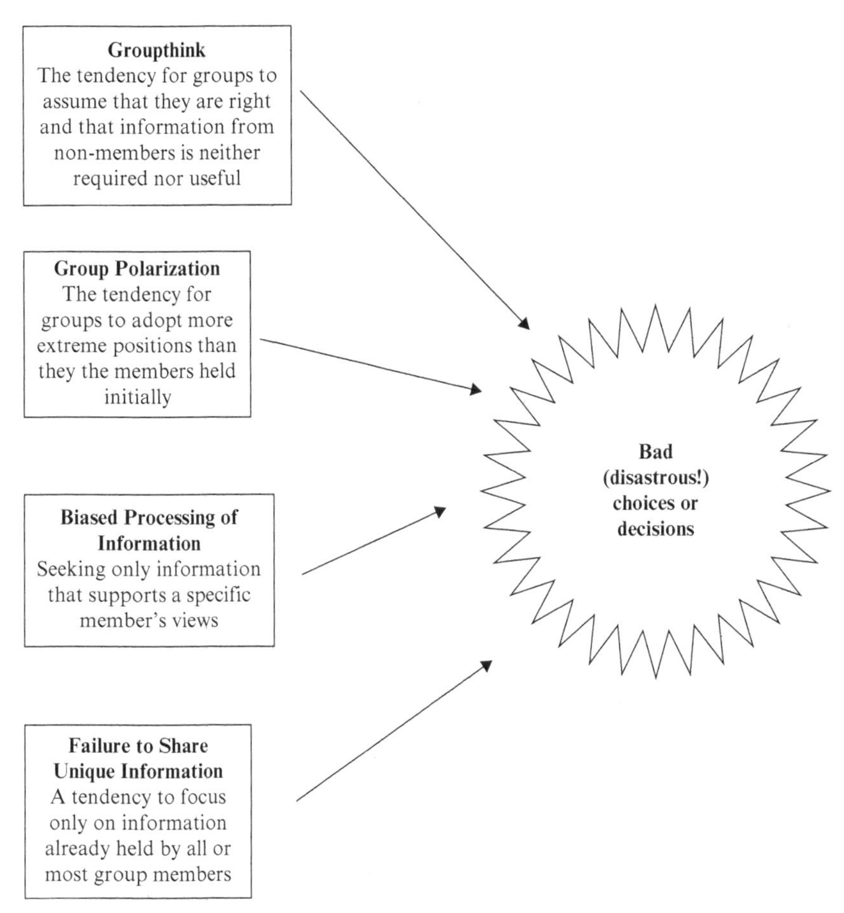

Note: As shown here, there are several processes—groupthink, group polarization, biased processing of information (e.g., focusing on being "right" rather than accurate), and failing to share information held by only some members.

Source: Robert A. Baron.

Figure 7.5 *Why groups sometimes make devastating decisions*

and our decisions. Signal detection theory suggests that in many situations involving decisions, a key task is deciding whether a stimulus is or is not really present. Entrepreneurs encounter this task in their efforts to recognize opportunities, and then determine if the opportunities are real. Signal detection theory suggests that in such instances, four possible outcomes are possible: the opportunity is really present and is recognized (this is termed a "hit"), it is not present, but entrepreneurs conclude that it is (a false alarm), it is not present and entrepreneurs correctly recognize this fact (a correct rejection), or it is present and they do not realize that it is (a miss). How well individuals do in performing this task depends on where they set their subjective criterion for concluding that a stimulus (e.g., an opportunity) is or is not present.

Groups are often called upon to make important decisions, largely because it is assumed that they offer many benefits (e.g., members can pool their skills and knowledge). In fact, however, existing evidence indicates that groups often make bad or even disastrous decisions. This can occur because of *group polarization*—a strong tendency for groups to shift, during discussions, to more extreme positions than those held by any members initially, the failure of group members to share information held uniquely by individual members, and a tendency to process information in a biased way that supports initial judgments and "being right" rather than accuracy in making decisions.

References

Baron, R.A. (2006). Opportunity recognition as pattern recognition: How entrepreneurs "connect the dots" to identify new business opportunities. *Academy of Management Perspectives*, 20, 104–119.

Beach, L.R., and Mitchell, T.R. (1990). Image theory: a behavioral theory of decisions in organizations. In Staw, B.M. and Cummings, L.L. (eds), *Research in Organizational Behavior*, Vol. 12. Greenwich, CT: JAI.

Busenitz, L.W., and Barney, J.B. (1997). Differences between entrepreneurs and managers in large organizations: Biases and heuristics in strategic decision-making. *Journal of Business Venturing*, 12, 9–30.

Dunegan, K.J. (1995). Image theory: Testing the role of image compatibility in progress decisions. *Organizational Behavior and Human Decision Processes*, 62, 79–86.

Gigone, E., and Hastie, R. (1997). The impact of information on small group choice. *Journal of Personality and Social Psychology*, 72, 132–140.

Greenberg, J., and Baron, R.A. (2009). *Behavior in Organizations*, 9th edn. New York: McGraw Hill.

Hardman, D. (2009). *Judgment and Decision Making*. London: British Psychological Association Press. Upper Saddle River, NJ: Pearson Education.

Hirt, E.R., and Markman, K.D. (1995), Multiple explanation: A consider-an-alternative strategy for debiasing judgments. *Journal of Personality and Social Psychology*, 69, 1069–1086.

Jarvstad, A., Rushtonk, S.K., Warren, P.A., and Hahn, U. (2012). Knowing when to move on: Cognitive and perceptual decisions in time. *Psychological Science*, 23, 589–597.

Kahneman, D. (2003). Maps of bounded rationality: psychology for behavioral economics. *The American Economic Review*, 93, 1449–1475.

Krueger, N.J., and Brazeal, D.H. (1994). Entrepreneurial potential and potential entrepreneurs. *Entrepreneurship Theory and Practice*, 18(3), 91–104.

Leotti, LA., and Delgado, M.R. (2012). The inherent reward of choice. *Psychological Science*, 23, 1310–1318.

Mojzisch, A., and Schulz-Hardt, S. (2011). Process gains in group decision-making: a conceptual analysis, preliminary data, and tools for practitioners. *Journal of Managerial Psychology*, 26, 235–246.

Schwartz, B. (2004). The tyranny of choice. *Scientific American*, April, 71–75.

Schwartz, B., Ward, A., Monterosso, J., Lyubomirsky, S., White, K., and Lehman, D.R. (2002). Maximizing versus satisficing: happiness is a matter of choice. *Journal of Personality and Social Psychology*, 83, 1178–1192.

Stewart. W.H., and Roth, P.L. (2001). Risk propensity differences between entrepreneurs and managers in large organizations: biases and heuristics in strategic decision-making managers: a meta-analytic review. *Journal of Applied Psychology*, 86, 145–153.

Swets, J.A. (1992). The science of choosing the right decision threshold in high-stakes diagnostics. *American Psychologist*, 47, 522–532.

8 Managing adversity: dealing with stress, learning from our mistakes, and coping with failure

Chapter outline

Stress: its basic nature, and how to manage it successfully
Stress: its devastating effects
Coping strategies: contrasting ways of dealing with stress
Why some people—including entrepreneurs—are better at handling stress
Learning from our mistakes: one reason why there *is* often strength in adversity
Imagining a better way: counterfactual thinking
How to give—and accept—criticism
Dealing with failure—and building personal well-being
Effects of business failure: financial, social, psychological
Learning from business failure—and bouncing back
Subjective well-being: a protective shield against the effects of adversity

* * *

What doesn't kill us makes us stronger.
(Friedrich Nietzsche)

I know I have a very tough five months ahead of me, but I will get through those months knowing that I have the ability to return to my productive . . . life, my interesting work and future business opportunities.
(Martha Stewart)

Failure is the condiment that gives success its flavor.
(Truman Capote)

Life's journey is anything but tranquil. Everyone experiences major "ups" and major "downs", and rare is the person whose entire life unfolds without major problems or troubles. Further, the more ambitious our goals, and the higher our expectations, the more likely we are to be frustrated by obstacles, experience disappointments, and come face-to-face with failure. These experiences are largely inevitable, so the capacity to deal with them effectively is one of the basic life skills everyone should acquire. Entrepreneurs, of course, are at very high risk for experiencing adversity: their goals are lofty, their aspirations bold, and their commitment to reaching them is often very powerful (e.g., Murnieks et al., in press, 2012). Further, the best paths for achieving these goals and realizing these aspirations are often uncertain—a fact that virtually guarantees many trials—and many errors—en route toward them.

For these reasons, developing the capacity to manage adversity effectively should be included in any list of the tools entrepreneurs need to achieve their central goals. In this chapter, we will examine evidence related to several aspects of this capability, and how to best develop it. First, we will focus on stress—what it is, the effects it produces, and most importantly, how it can be managed effectively. Next, we will turn to an intriguing and basic question: "How can individuals best learn from their mistakes?" Since errors and missteps are inevitable and will occur, it is important to convert these experiences into knowledge, and better performance in the future; but are there best ways to do so? Third, we will examine the sometimes devastating effects of failure, and how, perhaps, entrepreneurs can best handle this most ultimate form of adversity so that they arise from it stronger than before (as Nietzsche suggests above), rather than as battered or even totally shattered. As part of this discussion, we will consider the varied foundations of subjective well-being, another term for personal life satisfaction or happiness, because growing evidence indicates that building such satisfaction into our lives can provide a powerful shield against all forms of adversity, including failure.

Stress: its basic nature and how to manage it successfully

Have you ever felt right on the edge of being overwhelmed by events in your life or by pressures you could no longer handle? If so, you are already all too familiar with stress—a combination of physiological, emotional, and cognitive reactions that together, can seriously disrupt our physical and psychological functioning. To put is simply: at high levels, stress feels bad, involving such physical symptoms as sweaty palms, dry mouth, a racing heart, feelings of dread and anxiety (e.g., fear of pending failure), and thoughts of being unable to cope. These reactions are often produced by external events (various stressors), and in their efforts to launch and operate new ventures, and perhaps change the world in some way, entrepreneurs face a daunting array of such conditions. The environments in which they work are often unpredictable and subject to rapid change, their work load is heavy, they have responsibility for their companies and its employees, must make many decisions—often in the absence of essential information), and frequently operate under severe financial constraints. If they work in large organizations and want to generate change, they must often do so without the support and assistance from their co-workers or supervisors. A large body of evidence on the causes of stress (e.g., Xie et al., 2008; Ivancevich and Matteson, 1980; Jex and Beehr, 1991) suggests that as a result of these conditions, entrepreneurs would logically be expected to experience high levels of stress—levels greater than those experienced by persons in other careers (e.g., managers or employees).

While this may generally be true, we will soon see that it may actually be the case that entrepreneurs are especially able to deal with (i.e., cope with) such environments. Before examining that possibility—which is supported by recent research findings (e.g., Baron et al., 2012)—we will first examine the effects of stress and several ways of handling it successfully.

Stress: its devastating effects

Stress has sometimes been termed the "silent killer", and aptly so, because it is strongly related, in negative ways, to physical health. In fact, prolonged exposure to high levels of stress (i.e., exposure to many stressors), has been found to play a role in almost every major disease known to modern medicine, including heart disease, high blood pressure, hardening of the arteries, ulcers, and diabetes (e.g., Kiecolt-Glaser et al., 2001). How does stress produce such effects? While we do not yet know the precise mechanisms involved, a general model of how stress exerts adverse effects on personal health has emerged, and goes something like this. By draining our resources, inducing negative feelings, and keeping us off balance physiologically, stress upsets our complex internal chemistry. As a result, it interferes with efficient functioning of the immune system—the mechanism that allows our bodies to identify and destroy potentially harmful microbes and substances, such bacteria, viruses, and our own cells when they become cancerous. Our bodies are, in a sense, a totally "open system"—every breath we take, and every mouthful we swallow can introduce potentially harmful agents. An effective immune system, then, is the primary defense we have in this continuous battle; and in general, the immune system does its job incredibly well. Every day, it removes or destroys many potential threats to our health.

Prolonged exposure to high levels of stress, however, disrupts this system. For instance, prolonged exposure to stress reduces the level of lymphocytes (white blood cells that fight infection and disease), and increases levels of the hormone cortisol (hydrocortisone), a substance produced by our own bodies that has beneficial functions such as raising blood sugar levels when energy is needed, but also suppresses the immune system. In addition to these effects, stress adversely affects personal health indirectly by influencing health-related behaviors (see Figure 8.1). If you have ever

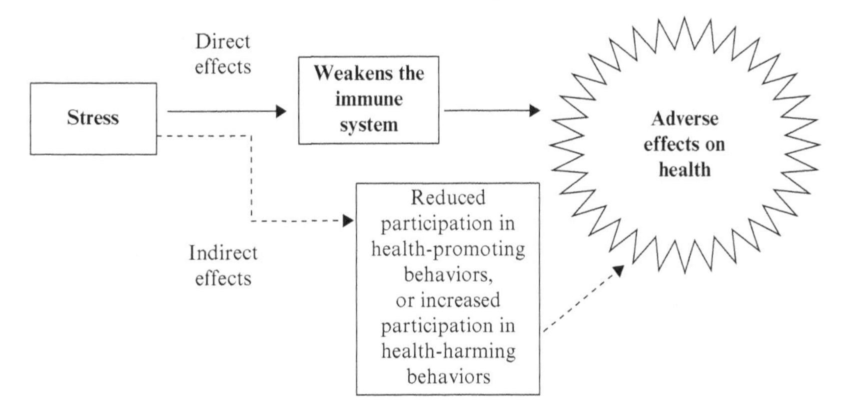

Note: As shown here, stress affects personal health directly, by weakening the immune system. In addition, it also affects indirectly by interfering with health-promoting behaviors, or encouraging health-harming actions.

Source: Robert A. Baron.

Figure 8.1 *How does stress affect personal health?*

felt so "stressed out" that going to the gym seemed to be too much effort, or indulged yourself in high-calorie snacks (or several alcoholic drinks) to offset a bad mood, you are familiar with such indirect effects. In other words, some of the means we use to cope with stress can succeed in the short term—they do help us feel better—but can undermine our health in the long term. Regardless of the precise mechanisms involved, however, it is clear that stress definitely has harmful effects on physical health. It also has detrimental effects on our psychological well-being, generating negative feelings, anxiety, and a syndrome known as burn-out—strong feelings of cynicism about the value of one's work or career coupled with loss of interest in and enthusiasm for it (e.g., Berglas, 2001). In other words, stress can, under some conditions, generate effects precisely opposite to those described by the term passion (see discussion of this topic in Chapter 6).

Coping strategies: contrasting ways of dealing with stress

Stress is an inescapable aspect of life, and as we have already noted, entrepreneurs, because of their efforts to turn ideas into reality, seem especially likely to experience it. A key question then, is what are the best strategies for managing stress when it occurs? Several exist, but they vary greatly in terms of their effectiveness. In the discussion below, we will move from strategies that are highly effective to ones are less effective (e.g., Chao, 2011; Lazarus and Folkman, 1984).

Problem-focused Coping: One highly effective means of coping with stress is to do something about its causes—solve problems causing stress, develop plans for reducing it. By reducing the underlying causes of stress, stress itself is decreased—and if the solutions are good ones, they are reduced in a long-term manner.

Seeking Social Support: Another effective tactic is to seek help, guidance, and support from others. They can often suggest ways of reducing stress, and the counsel and support they provide can reduce its harmful effects.

Distancing: This strategy involves ignoring stress or concluding that it is tolerable. Clearly, this does nothing to reduce the causes of stress or to deter its harmful effects.

Avoidant Coping: In some ways, this is not only ineffective—it is also potentially dangerous. Avoidant coping involves such actions as wishing the stress would vanish or "venting" negative emotions. This, once again, does nothing to address or reduce the causes of stress.

Short-Term Tactics for Reducing Negative Feelings: Stress, as noted earlier, feels bad, and this sometimes leads individuals to engage in actions that counter such feelings in the short-run: substance abuse, gambling, binge eating, unrestrained sexual adventures. Clearly, these are counter-productive and can result in much more harm than good.

Even these brief descriptions suggest strongly that some means of coping with stress are much better than others. When entrepreneurs experience high levels of stress, therefore, they should—like everyone else—focus on addressing the causes directly (problem-focused coping), and seeking social support. The other tactics listed above are much less likely to produce lasting relief from stress and other beneficial outcomes. Fortunately, although entrepreneurs live in worlds filled with stressors, they do have a greater degree of control over their own schedules and actions, and do not have to follow the orders or directives of supervisors. This suggests although they are exposed to many sources of stress, they may also have better opportunities to deal with them than persons who are not self-employed. Whether they take advantage of these opportunities, however, is also in their own hands!

Why some people—including entrepreneurs—are better at handling stress

While most people find stress to be unpleasant, others seem to thrive in its presence. For instance, as described in Chapter 1, emergency room physicians, flight traffic controllers, professional acrobats, and people in many other careers or professions seem to actually enjoy conditions most persons would find highly stressful and unpleasant. What accounts for this fact? One answer is that just as they differ in countless other ways, individuals differ in their capacity to endure stress. Several factors have been found to play a role in this respect but among these, one that is highly relevant to entrepreneurship is known as *psychological capital* (Peterson et al., 2011).

Psychological capital, which like human or social capital can range from low to high, involves several components: high levels of self-efficacy, optimism, hope, and resilience (Luthans et al., 2005). As defined by Luthans et al. (2007, p. 3), these factors refer to:

> (a) having confidence (efficacy) to take on challenging tasks and put in the necessary effort to succeed at them; (b) persevering toward goals and, when necessary, redirecting paths to goals (hope) in order to succeed; (c) making a positive attribution (optimism) about succeeding now and in the future; and (d) when beset by problems and adversity, sustaining and bouncing back and even beyond (resilience) to attain success.

As you can see, these are related to several skills or capacities we have considered in previous chapters, including self-efficacy, optimism, willingness to improvise. However, together, they seem to form a uniquely beneficial combination. Recent evidence indicates that psychological capital is related to several important outcomes in work settings: superior performance, more positive work-related attitudes (e.g., job satisfaction, organizational commitment), and lower turnover (Peterson et al., 2011). In addition, other findings indicate that psychological capital is negatively related to stress. For example, Avey et al. (2009) found that among a large sample of employed persons, the higher their psychological capital the fewer symptoms of stress they experienced. Similarly, a meta-analysis of research on the impact of psychological capital (Avey et al., 2011) reported that psychological capital was negatively related to job stress as well as undesirable employee behaviors (i.e., various forms of work-place deviance such as employee theft).

Together, these findings suggest that psychological capital can provide an effective buffer against high levels of stress. And in fact, one recent study (Baron et al., 2012), found that among a large sample of entrepreneurs, the higher their psychological capital (as measured by a short questionnaire designed to assess this variable), the less stress they reported experiencing and the higher their feelings of subjective well-being. (We will return to subjective well-being in detail in a later section, but in general, it refers to the extent to which individuals are satisfied with their current lives; Diener and Chan, 2011.) In addition, the higher the entrepreneurs' psychological capital, the higher the performance of their companies, at least as described by the entrepreneurs themselves. Another finding of interest was that the level of stress reported by these entrepreneurs was actually lower than that reported on a standard measure of stress than any other occupational group (Cohen and Janicki-Deverts, 2012). Overall, results offered support for the model shown in Figure 8.2—one in which

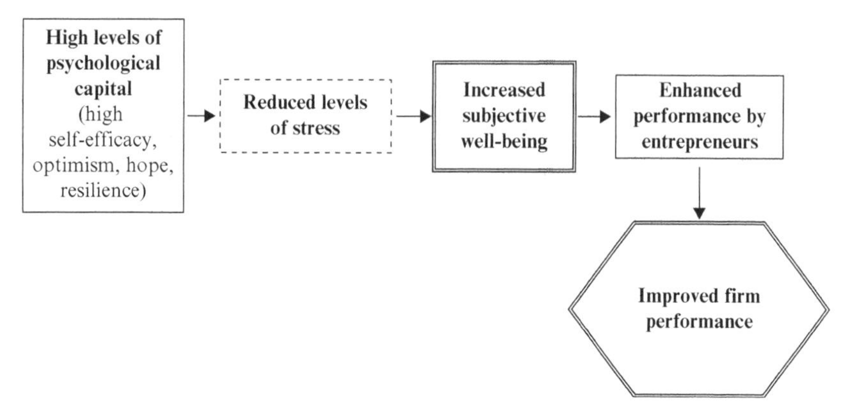

Note: Recent findings indicate that high levels of psychological capital reduce the levels of stress experienced by entrepreneurs, enhance both their subjective well-being and, indirectly, increase the success of their new ventures.

Source: Based on findings reported by Baron et al., 2012.

Figure 8.2 *The benefits of psychological capital*

high levels of psychological capital reduce the levels of stress experienced by entrepreneurs, and so enhance both their personal well-being and the success of their new ventures. To the extent this model is verified in further research, it suggests that developing the components that combine to produce psychological capital can be very beneficial for entrepreneurs—or, perhaps, anyone else exposed to high levels of unavoidable stress.

Why do entrepreneurs report relatively low levels of stress? Before concluding this discussion, it is important to comment briefly on the finding that despite their exposure to many stressors, entrepreneurs appear to report relatively low levels of stress (Baron et al., 2012). Although that might at first seem puzzling, it is actually predicted by the attraction-attrition-selection model described in Chapter 1 (e.g., Schneider, 1987). As you may recall, only some individuals find entrepreneurship attractive as a potential career, only some of these discover that it is really to their liking, and only some survive in this role. In other words, perhaps entrepreneurs are individuals who are selected, both by their own preferences and the requirements of the entrepreneurial role, to be persons who are relatively high in certain characteristics (e.g., the capacity to tolerate high levels of uncertainty and stress; self-confidence, optimism, the desire for personal autonomy, etc.). If that is true, then it is not at all surprising that they would find conditions other people perceive as highly stressful to be lower on this dimension. At present, this is only an interesting possibility, but it does fit well with evidence indicating that because of the process of attraction-attrition-selection, persons in various occupations do indeed differ in many ways.

Learning from our mistakes: one reason why there *is* often strength in adversity

George Washington once observed that "To rectify past blunders is impossible, but we should profit by the experience of them." In this sentiment, he agreed with countless keen observers of human behavior—and human nature—who have pointed out that being imperfect, we all make errors, but as long as we learn from these mistakes, they are well worthwhile. Clearly, this sentiment is true for entrepreneurs. Since they are attempting to do something truly new, they will almost certainly make many mistakes; that would seem to be an inescapable part of the process. An important, and very practical question, then, is how can they best profit from these mistakes? How can they effectively use them as a source of new knowledge and, perhaps, improved future performance? Learning is a general process that permeates all human activities, and psychologists generally define it as any relatively permanent change in behavior or thought resulting from experience (e.g., Domjan, 2003). Here, however, we will focus on two factors that can strongly contribute to the process of learning from our mistakes.

Imagining a better way: counterfactual thinking

Life, as we have noted several times (and you already know very well!), is filled with disappointments. All too often, things do not turn out the way we hoped. When this occurs, an intriguing cognitive process often operates—a process known as *counterfactual thinking* (e.g., Roese, 1997). This process involves "imagining what might have been"—that is, imagining outcomes different from the ones that actually occurred. Did we fail to get a job we badly wanted? We may entertain thoughts about what it would have been like to obtain it. Did venture capitalists turn down a request for financial support? Entrepreneurs may imagine what might have happened if they had said "yes".

The tendency to imagine outcomes and events different from the ones that actually occurred is very strong, and occurs in many situations. Often, counterfactual thinking involves imagining better outcomes than actually occurred (upward counterfactual thinking); but in other cases, it involves imagining worse outcomes (downward counterfactual thinking). Here is a vivid illustration of this fact. Among winners of Olympic medals, which group do you think is happiest? Those who receive gold, those who receive silver, or those who receive bronze? While you might guess that those receiving the top prize (gold) are happiest, in fact, it is the athletes who win bronze medals who are happiest. Why? Because they can readily imagine what it would have been like to have received no medal at all; they engage in downward counterfactual thinking. And it is the silver-medal winners who are least happy: they can readily imagine winning gold, and doing so leads to strong feelings of disappointment and regret (Medvec et al., 1995).

Overall, engaging in counterfactual thinking offers a mixed picture of benefits and costs. On the negative side are strong feelings of regret ("If only I had . . ."), and other strong negative thoughts and reactions. These are accentuated when individuals feel unable to do better in the future (Sanna, 1997). On the positive side are positive feelings induced when individuals compare their current outcomes with ones that are worse, as in the case of bronze Olympic-medal winners. Even more important, engaging in counterfactual thinking can often help mitigate the bitterness of disappointments. After tragic events, such as the death of a loved one or a devastating natural disaster, many people find solace in such thoughts as "Nothing more could have been done; his/her death was inevitable," or "This was an act of nature—it could not have been prevented." On the other hand, imagining better outcomes with respect to these events —"If only the illness had been diagnosed sooner" or "If only that new dam had been built" can actually intensify the sadness and grief they produce.

Most relevant to the present discussion about learning from our mistakes, however, is an additional effect of counterfactual thinking. When imagining different—and usually better outcomes—individuals may also be encouraged to ask themselves, "What could I have done differently—or better? How could I have prepared better, or used my time more effectively?" (e.g., Kray et al., 2006). This can be very beneficial in terms of turning disappointments and setbacks into learning experiences—events that provide important clues about how to actually do better in the future. Many highly successful entrepreneurs report that they learned a great deal from their early setbacks, and they further indicate that this is so because of the effort they invested in thinking about what they did wrong, and how they could avoid such errors in the future. On the other hand, if individuals simply conjure up images of success and achievement without trying to understand *why* they failed in reality, counterfactual thinking can actually interfere with learning and developing the skills and strategies needed for actual success. (Interestingly, because they are so "future-oriented", it appears that entrepreneurs, as a group, engage in counterfactual thinking less often than other persons; Baron, 2000.)

In sum, counterfactual thinking is indeed something of a "two-edged sword". If used effectively, it can diminish the pain of disappointments, and simultaneously point the way toward attaining better outcomes. If, instead, it focuses on actions or decisions not taken in the past, and that cannot now be changed, it can lead to strong feelings of regret that are both futile and emotionally painful. The key, then, is to use this process as a guide to actually doing better, rather than a source of daydreams about

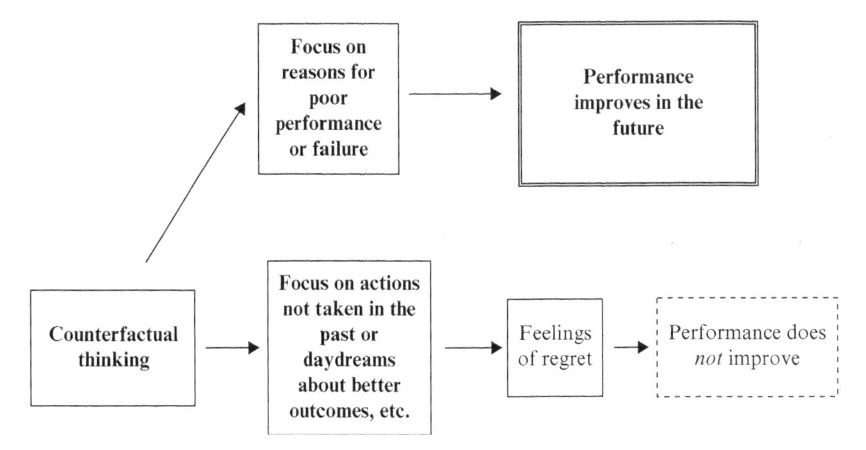

Note: Counterfactual thinking—imagining what "might have been"—can enhance learning from previous mistakes *if* it focuses on understanding the reasons for poor performance or failure. If, instead, it focuses on actions that were not taken in the past, it leads mainly to feelings of regret and will not facilitate doing better in the future.

Source: Robert A. Baron.

Figure 8.3 *Counterfactual thinking and learning from our mistakes*

what we hoped would happen but did not. (Please see Figure 8.3 for a summary of these points.)

How to give—and accept—criticism

In an earlier chapter (Chapter 6), we noted that willingness to seek, and accept advice from other persons—especially ones who are skilled or competent in a particular domain—is an important ingredient in success for entrepreneurs. This is eminently reasonable, and in fact, many people do ask for help when they need it—especially if they have a good understanding of what they know and what they do not know (see Chapter 4). Advice is, in a sense, a specific kind of feedback—feedback that is requested. Feedback, it is widely recognized, is essential in situations where individuals genuinely wish to improve in some way, but unfortunately, many people are not very good either at providing it or accepting and using it when they receive it. This is especially true of negative feedback—often termed *criticism*. Such feedback can be extremely helpful in facilitating the process of learning from our mistakes, because often it is easier for other persons to recognize what we are doing wrong and how we can do better than it is to acquire this information for ourselves. This is one reason why "personal coaches" are now so popular—they provide the detailed feedback that can help even highly skilled persons such as surgeons—improve (e.g., Gawande, 2011). But research findings indicate that the style in which such feedback is delivered is crucial in determining whether it is accepted or rejected as unfair and useless.

Basically, negative feedback—especially informal negative feedback—can be delivered in two forms. One, described as *destructive criticism* (e.g., Baron, 1990, 1993), is, sadly, the form that is often delivered. Such feedback is harsh and often sarcastic. It contains threats ("If you don't improve, you are out of here!"), occurs long after the performance being criticized, and puts the blame for poor performance squarely on the shoulders of the recipient. It is general rather than specific in nature—that is, it does not clearly identify what was wrong or should be improved, and does not provide clear suggestions for achieving better performance. Such criticism is all too common—much more common than the other form, *constructive criticism*, which is the opposite in all these respects. Such criticism is considerate and reasonable, contains no threats, occurs soon after the performance being evaluated, does not blame the recipient for poor performance, is specific in nature, and focuses on performance, not the person being criticized. As you might expect, existing evidence indicates that constructive criticism is much more likely to be perceived as helpful by recipients, and to actually help them improve their performance (Baron, 1993).

Since entrepreneurs work with many different persons and are often in a leadership position (after all, it is they who have founded a company or are attempting to change the status quo), they must often deliver feedback to others—employees, their co-founders, suppliers. If they genuinely want these people to improve in the ways they identify, then delivering feedback constructively is essential. It is crucial, then, for entrepreneurs to ask themselves the following question *before* they offer negative feedback: "Why am I doing this?" The only rational answer is "Because I want to help this person improve—do better." Anything else—any other motives—(e.g., "I'm getting even", "I'm teaching her/him his place", "I want to show who's in charge here") is not only inappropriate—it is likely to be counterproductive too.

With practice, most people can master the art of delivering constructive rather than destructive criticism. More difficult, however, is learning to accept negative feedback from others. No one enjoys being told that they are not doing well, and reactions to such information, even if it is delivered constructively, range from disbelief, through open rejection, and then to anger and resentment. These feelings are understandable, but as noted by Chris Argyris (2002), a highly insightful observer of human behavior in business contexts, they get in the way of effective learning. If we reject criticism from others, how can we benefit from it and actually improve whatever we do? Even worse, as Argyris notes, it is precisely the people we would expect to be most open to such feedback who are least likely to accept it. On the basis of more than a decade of research, Argyris (2002) has reached the conclusion that the very best people in most organizations are people who, in essence, have rarely experienced failure. They were A+ students in school, won very good starting jobs, and have moved upward, from success to success ever since. As a result, when they do receive negative feedback or experience failures, they have little idea about how to learn from it. Rather, they tend to become defensive, attributing such outcomes to external forces—the incompetence of other persons, the unreasonableness of customers—and assign little or no responsibility for these disappointments to themselves. Argyris describes this is as "double doom loop". Not only do they attribute disappointing results to others, they feel threatened by such results and by any suggestion that they themselves are to blame. The outcome is predictable: their ability to learn from their mistakes is seriously compromised, so that that they do not improve, even in the face of valid and constructively delivered negative feedback.

Interestingly, it is often the individuals most in need of feedback—and especially criticism—that reject it most (Dunning, 2005). Specifically, poor performers in almost any field tend to overestimate their own effectiveness or success to a greater extent than high performers. In college classes, for example, the students who score lowest on exams often predict (before receiving their grades) that their scores will be high—higher, in fact, than those predicted by the students who actually receive the highest scores (e.g., Dunning et al., 2003). This occurs because such persons lack insight into their own poor skills—in other words, they are lacking in understanding of "what they know and do not know", as discussed in Chapter 4).

The moral in all this for entrepreneurs seems clear: how you deliver negative feedback to others, and how you handle it when you are on the receiving end, are crucial from the point of view of helping others learn from their mistakes—and accomplishing such learning ourselves. And since learning is a key foundation of excellent performance, these skills are ones entrepreneurs should seek to develop, and that should definitely be included among the tools they need for attaining success.

Dealing with failure—and building personal well-being

Adversity and setbacks are one thing; business failure is another. Given that a large percentage of all new ventures (and by extension, all attempts to act entrepreneurially) do indeed fail (or at least, come to an unsatisfactory end from the point of view of the entrepreneurs) it is important for entrepreneurs to understand the causes of such devastating outcomes as well as their effects. In a sense, every chapter in this book addresses these issues. The basic logic, throughout, has been that although there are many reasons why entrepreneurs fail in their efforts to convert their ideas into tangible and beneficial outcomes, very important among these is their lack of various tools (skills, motives, interests, etc.) that they need for achieving success. We will review these tools—which have been the key focus of the present book—in detail in Chapter 9, but here, we want to note that even if entrepreneurs do possess the tools they need, they may still experience failure (although the likelihood of this outcome is much reduced). In this discussion, we will define business failure as it is defined by Ucbasaran et al. (in press, 2012): *the cessation of a new venture because it has failed to meet a minimum threshold for economic or other viability adopted by the entrepreneur*. In other words, we will focus here only on those instances in which a company ceases operations because it has not been successful in developing the opportunities on which it is based—in essence, it has not made sufficient profits to stay in business. There are many other reasons why a new business may cease to exist—for instance, sale to another person or company, a desire on the part of the entrepreneur to "move on" to something else; but those are distinct from circumstances above to which the term "failure" does apply.

Effects of business failure: financial, social, psychological

The quotations at the start of this chapter suggest that failure can have important benefits. As we noted earlier, though, these involve learning from such experiences so that the "next time" will be better. And many successful entrepreneurs do note that they learned a great deal from their initial failures. They do not usually comment on the potential "downside" of such failures, however. As noted by Ucbasaran et al. (in press, 2012), these fall into three different categories: financial, social, psychological.

The financial costs of business failure are obvious: capital, time, and energy are all lost. If the entrepreneurs have used their own resources, the effects on their personal lives and finances may be ones from which they cannot readily recover. The social costs of business failure involve the loss of reputation and in extreme cases may result in the entrepreneur being labeled "a loser" to whom no one should listen or give support (i.e., the entrepreneur is actively stigmatized). The greater the extent to which entrepreneurs are blamed for business failures, the greater such social costs (e.g., Cardon et al., 2011). Overall, the loss of reputation associated with failures may make it more difficult for entrepreneurs to start again, even if they wish to do so.

Psychological costs of business failure, too, can be high, and include strong negative emotions (guilt, anger, shame), reductions in motivation (e.g., feelings of "helplessness"), and high levels of stress. Such effects are weaker for entrepreneurs who have previously run successful businesses

(Cope, 2011), among those who are high in confidence or self-efficacy (e.g., Hayward et al., 2010), and those who are high in emotional intelligence and can better manage their negative emotions (Shepherd, 2009). Of course, these financial, social, and psychological costs often combine and interact, so that, for instance, the financial losses intensify the emotional ones, and the psychological costs, in turn, may further negatively affect the entrepreneur's reputation. Overall, business failure is a bitter and disruptive experience for entrepreneurs, often—given their strong commitment to their new companies—among the most unpleasant of their lives.

Learning from business failure—and bouncing back

Failure is indeed a painful experience, but one thing that can make it at least a bit more tolerable is learning important lessons from it. As we noted earlier, many entrepreneurs believe that early setbacks and failures equipped them with the knowledge and skills they needed for later success. This is a compelling idea, but evidence about its accuracy is mixed, with some studies indicating that entrepreneurs do get better at important aspects of creating something new (e.g., recognizing opportunities; Ucbasaran et al., 2009), while others report that the companies started by entrepreneurs who have experienced failure are not necessarily more successful than those started by ones who have not experienced the bitter taste of failure (Ucbasaran et al., 2006). Overall, it appears that whether entrepreneurs do indeed learn from their mistake depends on two additional factors: how they explain such failures, and the emotions they experience as a result of failure.

Explanation for business failures involve, in part, attributions—the perceived causes of an event. To the extent entrepreneurs attribute failure to external causes—events beyond their control or the actions of other persons, learning is reduced (e.g., Ucbasaran, in press). To the extent they attribute failure to internal causes—their own actions, motivation, effort, characteristics, and mistakes!—the more likely they are to engage in counterfactual thinking, and hence to learn from their errors.

Similarly, intense emotions often interfere with learning, so to the extent the entrepreneurs experience intense negative feelings, the less likely they are to benefit from their failures. In a sense, they are too occupied with the task of coping with these emotions to process, store, and later use information related to the failure.

Bouncing back?

A key question that arises with respect to business failure is this: how do entrepreneurs attempt to cope with this experience? The better they do, the more likely they are to show resilience, and recover from such events. Revealing insights into these issues are provided by a growing body of evidence and theory concerning what is known as *grief recovery* (e.g., Shepherd, 2003, 2009; Shepherd et al., 2009). Basically, this work suggests that after experiencing failure of their businesses, entrepreneurs—like other persons—engage in several different tactics for coping with the intense negative feelings (grief) they experience.

One such strategy involves what is known as *loss-oriented tactics*—approaches in which they think deeply about the loss, trying, in general, to understand why it occurred, and an accompanying process known as *sensemaking*—which involves attaching meaning to events—efforts to "make sense" out of them (e.g., Weick, 1979). Another strategy involves *restoration-oriented tactics*—ones involving efforts to restore more positive moods or feelings through distraction or other techniques that shift attention away from the loss and the negative feelings it generates.

According to a model proposed by Shepherd (2009), the length of time required to recover from failure-generated grief can be reduced by certain steps: (1) scanning the environment for possible causes of the failure, (2) carefully interpreting this information, and (3) seeking to learn from these efforts. In short, cognitive mechanisms play a key role in determining how quickly—and how effectively—entrepreneurs recover from the intense negative emotions they experience as a result of failure. A general conclusion, then, is that entrepreneurs can indeed profit from failure experiences, but only if they take active steps to convert these unhappy events into genuine opportunities for learning.

Subjective well-being: a protective shield against the effects of adversity

So far, this chapter has focused on what might be termed "the negative side" of entrepreneurship: stress, disappointments, failure. Before concluding, then, we will restore the balance by examining precisely the opposite: the nature, sources, and benefits of human happiness—what is generally described as *subjective well-being* (Diener, 2000). Basically, this term refers to the extent to which individuals are satisfied with their lives—their responses to the following question: "How satisfied or dissatisfied are you with your life overall?" Careful research on this topic has yielded some very surprising results. For example, people appear to be happier than you might guess: all around the globe, in poor countries as well as rich ones they generally report very high levels of subjective well-being. In fact, about 80 percent of all respondents in large world-wide surveys indicate that they are very happy and satisfied with their lives (Diener et al., 2006). Perhaps even bigger surprises occur with respect to the sources of personal happiness. On the one hand, some of the factors that generate happiness are ones we would expect: for instance, having close friends, family, and loving spouses or romantic partners. Similarly, having specific goals and the resources to reach them, is also strongly linked to personal happiness (e.g., Diener, 2000). But a large surprise exists with respect to another factor many people believe is closely related to happiness: personal wealth.

Does wealth buy—or at least, enhance—happiness? Many people believe that the answer to this question is "Yes", and as we noted in Chapter 2, it is widely assumed that personal wealth is *the* central goal that leads individuals to become entrepreneurs—a belief that has been found to be inaccurate. But what about wealth itself? Wealthy persons can have almost anything they wish, whenever they want it, so shouldn't they be happy—or at least, happier than people who are less wealthy? Surprisingly, they are not. Existing evidence indicates that although there is some connection between wealth and happiness, this link is far less powerful than most people assume (e.g., Diener et al., 2011). Around the globe, household income is related to global feelings of well-being, and so is gross national product per capita, but primarily at low income levels, and only weakly. These findings suggest that once people have all the necessities, additional wealth above this point does not necessarily make them happier. This is clear at the national level: in research on 154 different countries, the links between gross domestic product (a measure of national wealth) and personal income on the one hand, and subjective well-being on the other do exist, but they are much weaker than the links between personal happiness and living in places that have good schools, good public transportation, and good health care (what are known as "public goods"). In fact, in many cases, persons living in countries that are relatively poor report higher levels of happiness than those in wealthier countries (see Table 8.1).

Why doesn't wealth necessarily generate high levels of subjective well-being? Several factors help explain these seemingly paradoxical findings. First, it appears that subjective well-being is more strongly connected with relative than absolute wealth. In other words, people seem to care more about how their income (wealth) compares with that of others than they do about the wealth itself (Boyce et al., 2010). In fact, other findings indicate that the greater the income inequality in a country at a given time, the lower the happiness people report (Oishi et al., 2011). Second, although wealth can provide desired possessions and many comforts, it often does not provide more time to enjoy them; in fact, wealthier people often report having less time to enjoy themselves than those who are less wealthy—they are too busy acquiring riches (Quoidbach et al., 2010). Finally, happiness seems to reside not in having what we want, but in wanting what we have—savoring the benefits we do have rather than focusing on the ones we do not (Larsen and McKibban, 2008). Whatever the precise reason, it seems clear that wealth, in and of itself, does not produce happiness. That is an important point for entrepreneurs to consider when asking themselves why they are attempting to create something new and better. Are they primarily

Table 8.1 *Does wealth automatically generate happiness?*

Country	*GDP Rank*	*Index of Subjective Well-Being*
United States	1	19
Denmark	2	13
Japan	14	50
South Korea	24	84
Russia	36	72
Mexico	39	22
Ghana	68	51

Note: Growing evidence indicates that although wealth (as measured by gross domestic product) is related to subjective well-being, the relationship is much weaker than generally believed. Once people obtain the basic necessities of life, additional income does not boost their happiness. Rather, factors aside from wealth appear to be more important. As shown here, then, the wealthiest countries in financial terms are not necessarily the ones in which citizens express the greatest happiness. For instance, Japan is much wealthier than Ghana, but almost tied with it in terms of personal happiness.

Source: Based on data from Diener et al., 2010.

seeking financial gain (i.e., wealth)? Or are other goals—such as having a happy and stimulating life—equally, or even more important? The choice, of course, is largely theirs.

The many benefits of high levels of subjective well-being. Whatever its sources (and there appear to be many), a high level of subjective well-being offers important benefits. People high on this dimension experience positive moods and feelings more often, enjoy better personal relationships, attain more rapid promotions, express higher levels of job satisfaction (Wright et al., 2009), have better health (fewer and shorter illnesses; Lyubomirsky et al., 2005), and even live longer (!) than persons low in subjective well-being (Xu and Roberts, 2010). In addition—and most directly relevant to the present discussion—persons high in subjective well-being tolerate stress and setbacks better than those lower in personal life satisfaction (e.g., Baron et al., 2012). In short, being satisfied with one's life provides an important shield or buffer against the harmful effects of stress, which often does stem from adversity. That is certainly an important benefit that should not be overlooked!

Summary of key points

Adversity is a basic aspect of life, and entrepreneurs—who often set out to change the world in some way—are especially likely to experience it. Stress involves a combination of physiological, emotional, and cognitive reactions that together, can seriously disrupt our physical and psychological functioning and, subjectively, feel bad. Stress exerts harmful effects on both physical and psychological health, in part because it interferes with the effective functioning of the immune system and also because it can lead us to engage in behaviors harmful to our health. Several different tactics for coping with stress exist, including problem-focused coping, seeking social support, ignoring the stress, or engaging in actions that increase positive affect but may be dangerous (e.g., substance abuse). Clearly, tactics that seek to reduce or eliminate the causes of stress (e.g., problem-focused coping) are most effective in the long-term. Entrepreneurs differ in their capacity to deal effectively with stress, and this appears to stem, at least in part, from their psychological capital—a combination of self-efficacy, optimism, hope, and resilience. The higher entrepreneurs are in psychological capital, the lower the levels of stress they report, and the higher their subjective well-being. Developing a high level of psychological capital, then, can be very beneficial for entrepreneurs. Contrary to popular belief, entrepreneurs may experience lower levels of stress than persons in other fields or occupations, in part because only persons high in tolerance for stress become and remain entrepreneurs.

Mistakes are inevitable, but as long as they help us learn they can be well worthwhile. One process that can facilitate such learning is counterfactual thinking—imagining outcomes different from the ones that actually occurred. If individuals focus such thinking on how they could have done better in the past, this can help them learn from past errors, and so do better in the future. Another factor that can facilitate learning from past errors involves being able to both give and accept constructive criticism. Unfortunately, existing evidence indicates that it is the persons most in need of negative feedback that often reject it most strongly.

Business failure involves the cessation of a new venture because it has failed to meet a minimum threshold for economic or other criteria of success adopted by the entrepreneur. If it ends operations for other reasons (e.g., sale to another company or person), this does not constitute failure. Business failure has important financial, social, and psychological costs. Does experiencing such failure help entrepreneurs to be more successful in the future? Evidence on this point is mixed, but it appears that it can, if they focus on attempting to make sense out of this bitter experience, and—therefore—to learn from their own mistakes.

Contrary to popular belief, increasing wealth is not strongly related to subjective well-being—people's overall satisfaction with their lives. Some connection exists, but it is much weaker than was once believed. This appears to stem, at least in part, from the fact that people are more interested in the relative wealth than its absolute level, and that once they have attained a comfortable standard of living, additional wealth does not necessarily increase personal happiness. Whatever its specific sources a high level of subjective well-being offers many benefits, including enhanced resistance to the adverse effects of stress.

References

Argyris, C. (2002). Teaching smart people how to learn. *Harvard Business Review*, 4, 4–14.

Avey, J.B., Luthans, F., and Jensen, S.M. (2009). Psychological capital: A positive resource for combating employee stress and turnover. *Human Resource Management*, 48, 677–693.

Avey, J.B., Reichard, R.J., Luthans, F., and Mhatre, K.H. (2011). Meta-analysis of the impact of positive psychological capital on employee attitudes, behaviors, and performance. *Human Resource Development Quarterly*, 22(2), 127–152.

Baron, R.A. (1990). Countering the effects of destructive criticism: The relative efficacy of four potential interventions. *Journal of Applied Psychology*, 75, 235–245.

Baron, R.A. (1993). Criticism (informal negative feedback) as a source of perceived unfairness in organizations: effects, mechanisms, and countermeasures. In Cropanzano, R. (ed.), *Justice in the Workplace: Approaching fairness in human resource management*, pp. 155–170. Hillsdale, NJ: Erlbaum.

Baron, R.A. (2000). Counterfactual thinking and venture formation: The potential effects of thinking about "What might have been". *Journal of Business Venturing*, 15, 79–92.

Baron, R.A., Franklin, R.J., and Hmieleski, K.M. (2012). The stress-resistant entrepreneur: Why some entrepreneurs flourish while others shatter. Unpublished manuscript, currently under review.

Berglas, S. (2001). *Reclaiming the Fire: How successful people overcome burnout*. New York: Random House.

Boyce, D.J., Brown, G.D.A., and Moore, S.C. (2010). Money and happiness: rank of income, not income, affects life satisfaction. *Psychological Science*, 21, 471–475.

Cardon, M.S., Stevens, C.E., and Potter, D.R. (2011). Misfortunes or mistakes? Cultural sensemaking of entrepreneurial failure. *Journal of Business Venturing*, 26, 79–92.

Chao, R. (2011). Managing stress and maintaining well-being: Social support, problem-focused coping, and avoidant coping. *Journal of Counseling & Development*, 89(3), 338–348.

Cohen, S., and Janicki-Deverts, D. (2012). Who's stressed? Distributions of psychological stress in the United States in probability samples from 1983, 2006, and 2009. *Journal of Applied Social Psychology*, 42, 1320–1334.

Cope, J. (2011). Entrepreneurial learning from failure: an interpretative phenomenological analysis. *Journal of Business Venturing*, 26, 604–623.

Diener, E. (2000). Subjective well-being: the science of happiness and a proposal for a national index. *American Psychologist*, 55, 34–43.

Diener, E., and Chan, M. (2011). Happy people live longer: Subjective well-being contributes to health and longevity. *Applied Psychology: Health and Well-Being*, 3, 1–43.

Diener, E., Lucas, R., and Scollon, C.N. (2006). Beyond the hedonic treadmill. Revisiting the adaptation theory of well-being. *American Psychologist*, 61, 305–314.

Diener, E., Ng, W., Harter, J., and Arora, R. (2010). Wealth and happiness across the world: material prosperity predicts life evaluation, whereas psychosocial prosperity predicts positive feelings. *Journal of Personality and Social Psychology*, 99, 52–61.

Domjan, M. (2003). *Principles of Learning and Behavior*, 5th edn. Belmont, CA: Thomson/Wadsworth.

Dunning, D. (2005). *Self-insight: roadblocks and detours on the path to knowing thyself.* New York: Psychology Press.

Dunning, D., Johnson, K.I., Ehrlinger, J., and Kruger, J. (2003). Why people fail to recognize their own incompetence. *Current Directions in Psychological Science*, 12, 83–87.

Gawande, A. (2011). Personal bet. Top athletes and singers have coaches; should you? *New Yorker*, 3 October, pp. 63–72.

Hayward, M.A.L., Forster, W.R., Sarasvathy, S.D., and Fredrickson, B.L. (2010). Beyond hubris: how highly confident entrepreneurs rebound to venture again. *Journal of Business Venturing*, 25, 569–578.

Ivancevich, J., and Matteson, M. (1980). *Stress and Work: A managerial perspective*. Glenview, IL: Scott, Foresman and Company.

Jex, S., and Beehr, T. (1991). Emerging theoretical and methodological issues in the study of work-related stress. *Research in Personnel and Human Resource Management*, 9, 311–365.

Kiecolt-Glaser, J.K., Larsen, J.T., and McKibban, A.R. (2008). Is happiness having what you want, wanting what you have, or both? *Psychological Science*, 19, 371–377.

Kiecolt-Glaser, J.K., McGuire, L., Robles, T.F., and Glaser, R. (2001). Emotions, morbidity, and mortality: New Perspectives from Psychoneuroimmunology. *Annual Review of Psychology*, 53, 83–107.

Kray, L.J., Galinsky, A.D., and Wong, E. (2006). Thinking inside the box: the relational processing style elicited by counterfactual mind-sets. *Journal of Personality and Social Psychology*, 91, 33–48.

Larsen, J.T., and McKibban, A.R. (2008). Is happiness having what you want, wanting what you have, or both? *Psychological Science*, 19, 371–377.

Lazarus, R.A., and Folkman, S. (1984). *Stress, appraisal, and coping*. New York: Springer.

Luthans, F., Youssef, C.M., and Avolio, B.J. (2007). *Psychological Capital: Developing human competitive advantage*. New York: Oxford University Press.

Luthans, F., Avolio, B.J., Walumbwa, F.O., and Li, W. (2005). The psychological capital of Chinese workers: Exploring the relationship with performance. *Management and Organization Review*, 1(2), 249–271.

Lyubomirsky, S., Sheldon, K.M., and Schkade, D. (2005). Pursuing happiness: the architecture of sustainable change. *Review of General Psychology*, 9, 111–131.

Medvec, V.H., Madey, S.F., and Gilovich, T. (1995). When less is more: Counterfactual thinking and satisfaction among Olympic athletes. *Journal of Personality and Social Psychology*, 69, 601–610.

Murnieks, C.Y., Mosakowski, E., and Cardon, M.S. (in press, 2012). Pathways of passion: Identity centrality, passion, and behavior among entrepreneurs. *Journal of Management*.

Oishi, S., Kesebir, S., and Diener, E. (2011). Income inequality and happiness. *Psychological Science*, 2, 1095–1100.

Peterson, S.J., Luthans, F., Avolio, B.J., Walumbwa, F.O., and Zhang, Z. (2011). Psychological capital and employee performance: A latent growth modeling approach. *Personnel Psychology*, 64, 427–450.

Quoidbach, J., Dunn, E.W., Petrides, K.V. and Mikolajczak, M. (2010). Money giveth, money taken away: The dual effect of wealth on happiness. *Psychological Science*, 21, 759–763.

Roese, N.J. (1997). Counterfactual thinking. *Psychological Bulletin*, 121, 133–148.

Sanna, L.J. (1997). Self-efficacy and counterfactual thinking: Up a creek with and without a paddle. *Personality and Social Psychology Bulletin*, 654–666.

Schneider, B. (1987). The people make the place. *Personnel Psychology*, 40(3), 437–453.

Shepherd, D.A. (2003). Learning from business failure: Propositions of grief recovery for the self-employed. *Academy of Management Review*, 28(2), 318–328.

Shepherd, D.A., Wicklund, J. and Haynie, J.M. (2009). Grief recovery from the loss of a family business: A multi- and meso-level theory. *Journal of Business Venturing*, 24(2), 134–148.

Ucbasaran, D., Westhead, P., and Wright, M. (2006). *Habitual Entrepreneurs*. Cheltenham, UK and Northampton, MA, USA: Edward Elgar Publishing.

Ucbasaran, D., Westhead, P., and Wright, M. (2009). The extent and nature of opportunity identification by experienced entrepreneurs. *Journal of Business Venturing*, 24(2), 99–115.

Ucbasaran, D., Shepherd, D.A., Lockett, A., and Lyon, S.J. (in press). Life after business failure. The process of business failure for entrepreneurs. *Journal of Management*.

Ucbasaran, D., Westhead, P., Wright, M., and Flores, M. (2010). The nature of entrepreneurial experience, business failure and comparative optimism. *Journal of Business Venturing*, 25(6), 541–555.

Weick, C. (1979). *The Social Psychology of Organizing*. Reading, MA: Addison-Wesley.

Wright, T.A., Cropanzano, R., and Bonett, D.G. (2009). The role of employee psychological well-being in cardiovascular health: When the twain shall meet. *Journal of Organizational Behavior*, 30, 193–208.

Xie, J.L., Schaubroeck, J., and Lam, S.S.K. (2008). Theories of job stress and the role of traditional values: A longitudinal study in China. *Journal of Applied Psychology*, 93(4), 831–848.

Xu, J., and Roberts, R.E. (2010). The power of positive emotions: It's a matter of life or death—subjective well-being and longevity over 28 years in a general population. *Health Psychology*, 29, 9–19.

9 Putting it all together: a model of the highly effective entrepreneur

Chapter outline

Getting off to a very good start: from ideas to destinations
Motivation and goals: knowing what you seek—and why
Enhancing creativity—and avoiding "mental ruts"
Identifying opportunities worth pursuing
Tools for staying focused—and making progress
Self-knowledge: knowing what you know and do not
Working well with other people
Building a strong founding team
Developing effective social networks
Social and political skills: beyond social capital
Important steps along the way
Savor—but also manage—your enthusiasm and passion
Plan—but remain flexible
Making good decisions—alone, and in groups
Resilience in the face of the inevitable setbacks
Learning from mistakes
Personal happiness: the ultimate goal

* * *

> Effectiveness, in other words, is a habit; that is, a complex of practices. And practices can always be learned.
> (Peter Drucker)

> Nobody talks of entrepreneurship as survival, but that's exactly what it is and what nurtures creative thinking.
> (Anita Roddick)

> We should set our goals; then learn to control our appetites. Otherwise, we will lose ourselves in the confusion of the world.
> (Hark Herald Sarmiento)

At the very start of this book, it was noted that a large proportion of new ventures—and other entrepreneurial activities, too—fail, or at least disappear, within a short period of time. Given that entrepreneurs are usually deeply committed to their ideas and converting them into reality, and are willing to work long and hard to do so, this is a somewhat mysterious outcome. Why do so many experience failure? The basic thesis of this book suggests that although there are many (countless!) reasons for this seemingly paradoxical situation, central among them is the fact that despite their intense motivation and dedication, many *entrepreneurs lack certain tools essential for achieving success*. Describing these tools, and explaining how they can be acquired, has been the key goal of the entire volume. In this final chapter, this information will be combined into an integrated model of what might reasonably be termed *the highly capable or highly effective entrepreneur*. Such persons enter the challenging, risk-filled, competitive world of entrepreneurial activity with an array of skills, motives, interests, and characteristics that help them start well, make progress, stay focused, successfully accomplish key tasks, and bounce back after disappointments. In essence, their possession of these key tools (skill, characteristics, motives, interests, etc.) helps them to turn the odds in their favor, and to succeed when many others fail.

The remainder of this chapter will focus on these tools—the personal equipment that together help entrepreneurs succeed. ("Help" is the right term in this context because nothing, not even an ideal combination of personal skills, knowledge, and characteristics, can guarantee success in the uncertain and uncharted environments in which entrepreneurs often work.) Before beginning the central tasks of describing and summarizing these tools, however, I will offer a concrete example from my own life that illustrates this basic logic.

I have been an "angel investor" twice. The second time (which involved investing in a biotech start-up) was somewhat successful, but the first was a total disaster in which I, and every other investor, lost everything. This investment involved a company that not only seemed to offer great financial promise—it also closely matched my own strong, pro-environmental values. This company, called Cycletech, sought to develop a new way of recycling used automobile tires. The basic idea was to adapt new technology for freezing items to extremely low temperatures, for use with tires. At the target temperatures, the tires would become very brittle and could be easily fractured into a powder, from which valuable resources could be readily re-captured (nylon, steel, rubber, etc.). The technology was developed by scientists in Russia (the former Soviet Union), who were eager to emigrate to the United States and bring their knowledge with them. The process was based on sound scientific principles and findings and did seem to work: I watched demonstrations, and was very impressed.

Everything seemed to be moving along well: the company used the funds it received from investors to start construction on a new plant for freezing the tires, and purchased land on which the tires could be stored.

Note: Used tires are a blot on the landscape, so the idea of freezing them to temperatures at which they would shatter into a powder seemed very appealing. Valuable raw materials (steel, nylon, rubber) could then be recovered from this powder. Unfortunately, a company that sought to develop this idea—and in which the author personally invested—failed totally. This was not because the idea was a bad or impractical one, but, as explained to the author, occurred because the CEO of the company did not succeed in convincing government officials to grant the permits required for operations to proceed!

Source: Fotolia 35335886.

Figure 9.1 *Recycling used tires by freezing: a good idea but the wrong entrepreneur?*

The best part, perhaps, was that many towns were so eager to get rid of used tires, that they paid "tipping fees" to Cycletech to accept them. In other words, the raw materials were not only free—they were profitable in and of themselves! So, very soon, the company's acreage was filled with used tires (see Figure 9.1).

But then, major problems occurred: to operate, Cycletech needed many permits from the state government (it was located in New York, which was very strict about such facilities even twenty years ago). There seemed no reason why the company would not receive these—but in fact, it did not! Even worse, the relevant state agencies put the company on notice that it could no longer operate, and was in violation of many regulations. I was very puzzled—how could this be? And then, sadly, I learned the truth: the founder and CEO of the company somehow managed to anger the relevant state officials instead of obtaining their cooperation. The result? The company could not proceed, and ultimately, when it ran out of funds, went out of business, a total disaster for its investors.

It is hard for me to imagine a clearer illustration of the fact that even bright, hard-working, talented entrepreneurs often seem to pull "defeat from the jaws of victory". Why? It is argued throughout this book, that they fail, at least in part, because they are lacking in the skills or characteristics they need to succeed. Please note: external factors such as markets, actions of competitors, changes in technology, shifts in government policies, and many others are often *very* important. But the focus here is on entrepreneurs and their role in attaining the success they seek, so these factors will not be the primary focus of our attention.

One final point: please recall that all the statements and advice that follow are not simply reflections of my personal beliefs or ideas. Rather, they are, to the best of my ability, a reflection of the actual evidence now at our disposal concerning entrepreneurship—the process through which individuals use their own creativity, energy, skills and talents to develop something new, better, and beneficial and provide it to people who can use it (e.g., Baron, 2012). Having made these points and offered this example, let us now take a closer look at these key ingredients for entrepreneurial success.

Getting off to a very good start: from ideas to destinations

There is an old saying suggesting that "As you start, so you shall continue." The gist of this adage is that it is very important to start off strongly because a good start paves the way for continued success. That is certainly true with respect to entrepreneurship, and in this activity, "starting well" involves three basic components: motivation, creativity, and opportunities.

Motivation and goals: knowing what you seek—and why

One of the first tasks would-be entrepreneurs should undertake is that of identifying their own motives. They should be clear on *why* they have decided to act entrepreneurially—found a new company, try to change the way things are done in an existing one, or change the way in which they do their jobs or carry out their own career-related activities. Are they primarily interested in financial gains? Or are they motivated, instead, by the desires for independence, autonomy, recognition, or to grow as a person in some way? These contrasting motives do point to different paths and different actions, so it is crucial that potential entrepreneurs try to pinpoint their own key goals before they begin. If they wait, events and external pressures may coerce them into strategies or courses of action that are not consistent with what they truly seek, so the time to do this is at the very beginning.

Enhancing creativity—and avoiding "mental ruts"

Entrepreneurship, whatever its precise form, usually starts with an idea for something new and better, and that will generate desirable outcomes.

But how do such ideas emerge? In the past, it was widely believed that this process was, in a sense, "unknowable", and that the origins of ideas, especially good ones, were out of the realm of scientific knowledge. Decades of research on creativity, however, suggest that this is wrong: as we saw in Chapter 3, creativity, and the ideas it generates, involve basic cognitive processes, such as the key role of concepts (expanding them, combining them), and also attention to what is unusual or even obscure. It also involves having a large store of knowledge in a particular field or area, because it is impossible to combine or expand concepts or knowledge without first having them (Sternberg and Sternberg, 2011).

Perhaps most important of all, creativity—and the good ideas it generates—involves escaping from "mental ruts". We described one instance of such "ruts" in Chapter 3, where it was noted that Sony engineers, trapped by their own experience, developed early CDs that were 12 inches in diameter—the same size as long-playing records! Here is another: when the Spanish conquered the Inca (ruthlessly, unfortunately), they noticed that this otherwise advanced civilization did not use wheeled carts for transporting crops, building materials, or anything else. Yet, they also observed Inca children playing with toy carts that used the wheel in this way! When asked about this puzzling situation, Inca adults indicated that they had never thought of putting wheels on large carts. Why? They did not know, and when they saw the carts used by the Spanish, they immediately realized what a useful idea this was. Honestly, I cannot think of any clearer illustration of the powerful impact of "mental ruts" and how they can block creativity. The basic message for entrepreneurs is clear: to think creatively, it is necessary to cognitively escape from these ruts, and that can be accomplished by trying to imagine objects, processes, or almost anything else outside the context in which they are usually placed.

Identifying opportunities worth pursuing

Ideas for something new and useful are often the start of the process, but the next step, too, is crucial: determining whether, and how, such ideas can be used as the basis for opportunities—defined, in Chapter 3, as follows: perceived means of generating value (i.e., profit or other benefits) that are not currently being exploited and are perceived, in a given society, as desirable or, at least, socially acceptable. How can entrepreneurs best uncover (or create) these opportunities for action—for starting a new venture, proposing changes in existing companies, and so on? As we noted in earlier discussions, this involves several steps: actively searching for opportunities, identifying them by "connecting the dots" between several different factors or conditions, and then—crucially—determining whether the patterns (opportunities) that emerge are really feasible ones, worth pursuing.

Active search is, in fact, a skill that can be readily learned. One technique you can use to accomplish this is known as a "bug list". No, this term does not apply to identifying "pest" insects. Rather, it means being alert to things that "bug" or annoy you. Is there a product you use that does not work very well? Some aspect of your life that is less convenient than it could be? Some product that does not exist now but you wish did? Learning to actively think about such things can provide a good start toward identifying potentially beneficial opportunities.

In terms of "connecting the dots" (pattern recognition), this, too, is a skill that can be learned. In fact, it can be combined with your personal "bug list", because once you identify a problem or annoyance in daily life, you can try to find ways of reducing or solving it that, perhaps, do not now exist, but involve combining several seemingly unrelated factors. For instance, imagine that you have often been "bugged" by parking meters that accept only coins. If you do not happen to have enough change handy, you cannot park—or if you do, you run the risk of a parking ticket. What could be done to change this situation? Perhaps meters could be designed to accept credit cards, or some special card you purchase in advance from which various amounts could be deducted. This requires sophisticated technology, re-designed parking meters, and places to obtain the special keys or cards. Can these elements be combined into an opportunity? In fact, they already have; in many cities in the United States, parking meters that accept only coins are being replaced with ones that accept other means of payment, and so solve the "no coins" problem. This opportunity has already been recognized and exploited, but the general principle is the same for others you might personally identify.

Evaluating opportunities is, perhaps, the most difficult part of the process. But it is absolutely essential before beginning. It is all too easy to "fall in love with an idea" and assume, that since you like it so much, it must be one that will work. But this is where extra care is required: many ideas that seem practical and useful turn out to be unworkable for many reasons. For instance, the required technology may not be available, the costs of implementing the ideas may be higher than potential customers are willing to pay, or the market for them may be much smaller than it seems. It is truly crucial to examine such questions carefully before proceeding, and entrepreneurs who do—who temper their enthusiasm with information and facts—are much more likely to succeed than ones who do not.

In sum, getting off to a good start is crucial in entrepreneurship, just as it is in many other activities. But since entrepreneurship rests on creativity, ideas, innovations, and feasible opportunities (ones that can actually be developed), it is especially crucial in the process of making the "possible"—what we can imagine—real.

Tools for staying focused—and making progress

If you asked a large number of highly successful persons—athletes, scientists, executives, musicians, whatever—to describe the secrets of their success, you would receive many different answers. This is not surprising, because as we saw in Chapter 1, people are not very good at understanding their own motives and actions: why they did what they did in the past, why they experienced certain outcomes, and so on. But one theme you would probably hear over and over again involves *self-regulation*—regulating our own actions and feelings so as to enhance progress toward key goals. This can mean endless hours of practice, study, or effort, and it also implies refraining from actions that, although enjoyable, may interfere with or slow progress toward these goals. In other words, having effective *self-control* (what is often described as willpower) lies close to the heart of success.

Certainly, this is true for entrepreneurs. And it does not mean simply working many hours, not taking vacations, or even short breaks. Rather, it means exerting effort in highly focused, productive ways—working *smart*, not just long or hard. This is a difficult lesson for many entrepreneurs to grasp: they often boast about the very long hours they work. In fact, though, this is just as likely to mean that they are wasting a lot of time and effort as it is that they are making actual progress toward their goals. I have a friend who has started many new ventures, in different industries and activities, and he has been successful in almost all of them. Yet, when he has visited my classes he has often remarked: "If you are working more than 50 hours a week on your business, you are doing something wrong—you are not working 'smart'. The key is not the number of hours you invest, but what you do during that time." He often went on to note that staying focused (another important aspect of self-regulation, as described in Chapter 4) is also important. Two people can both spend 60 hours at work in their new ventures, but one may be focusing intently and persistently on key tasks relating to this business, while the other may be mixing such focus with other activities she or he enjoys—such as long conversations with partners (who are also likely to be friends), working on tasks she or he personally likes, while short-changing ones she or he does not like, but are also important. The results may be very different for these two entrepreneurs, despite their identical number of hours "at work". They just mean different things by the term "working". (If you have ever read the same paragraph over and over again, while you thoughts wandered, you know very well about the importance of focus in accomplishing many tasks!)

Self-knowledge: knowing what you know and do not

Self-control is a key aspect of self-regulation, but it is not the only one that is crucial. Another is what we described in Chapter 4 as *metacognition*—knowledge of, and control over, our own cognitive processes. Metacognition takes many different forms, but two are, perhaps, most important for entrepreneurs: (1) knowing what you know and do not know, and (2) recognizing when a current strategy or course of action is failing, and changing it or withdrawing from it.

Both of these are skills, or knowledge, that can be readily developed. Before beginning any task, it is very helpful to ask yourself: "Can I really do this? Do I have the knowledge and skill to succeed?" Answering truthfully requires a considerable dose of self-knowledge, but the benefits of developing this knowledge are immense. If you suffer from over-confidence (hubris, or ultra-high self-efficacy; see Chapter 6), you can quickly get into serious trouble. It is much better, if you conclude that you are "out of your depth" and to seek help rather than proceed on your own. Yes, many entrepreneurs are "women and men of action"—they want to forge ahead. But in many cases, it is much better to pause and carry out a careful self-assessment before proceeding.

Another major "trap" for entrepreneurs is sunk costs or escalation of commitment: continuing with a strategy or course of action when it yields negative outcomes. It is always difficult to know if this will change in the future, but it is very useful to set a "decision point" in such situations: "If losses reach this level, I will change directions no matter what" (see Figure 9.2). For instance, many investors (successful ones) often set such limits: "If a stock declines 20 percent, I'll sell it and take my losses, period—no questions, no quibbling." Again, that is doubly hard to do in entrepreneurial activities, but given that such activities are often carried out on a "shoestring" or with limited support from other persons in an organization or profession, establishing such "decision points" and sticking to them may truly help you avoid total disaster!

In sum, self-regulation involves many processes, but they are all ones that can be learned or developed. Effort spent in accomplishing these tasks can yield incredibly high dividends in terms of attaining the success you seek.

Working well with other people

As we noted in Chapter 5, entrepreneurs can rarely do it all themselves. Rather, they usually need help, cooperation, advice, and additional resources provided by other persons. For this reason, it is essential that

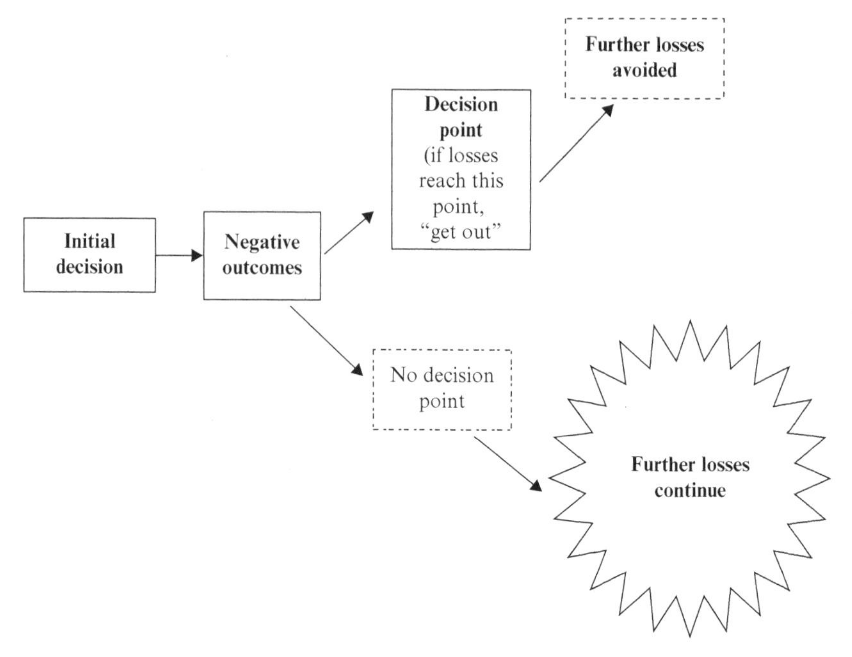

Note: It is all too easy for entrepreneurs to stick with a course of action even if it yields negative outcomes. Why? Partly because most people find it very difficult to admit that they made an error or bad decision, and to "cut their losses". Setting a decision point at which the decision to "get out" is made automatically—in advance—can be very helpful in avoiding this trap.

Source: Robert A. Baron.

Figure 9.2 *Avoiding escalation of commitment by setting a decision point*

they develop the skills useful for getting along with others, and for building strong and effective relationships with them. From choosing an outstanding founding team and attracting top-notch employees, through winning support from venture capitalists and securing orders from customers, these political or social skills are crucial. While many aspects of the "social side" of entrepreneurship are important, three are especially crucial: forming an excellent founding team, building strong social networks, and developing the social and political skills that facilitate these activities.

Building a strong founding team

As noted in Chapter 5, it is usually more comfortable or pleasant to work with people similar to ourselves than with people who are dissimilar. There are obvious benefits to forming a high-similarity (homogeneous) founding team—communication is improved, trust may be easier to develop, and agreement on key issues is more likely to occur. But—and this is an important "but"—it also virtually guarantees that the team will be lacking in certain key areas. For instance, a team consisting entirely of engineers will possess many technical skills relating to design and performance of products, but they may know very little about running a business—financial matters, marketing, human resource issues. This is why, in general, it is better—if possible—to assemble a founding team that is more heterogeneous. Ideally, the team should include people with expertise in all important areas. While this may be difficult to obtain, the closer to this goal entrepreneurs can come, the stronger their founding team—and new venture—will be. So be sure to give this matter careful thought and do you best to balance the comfort of team homogeneity against the benefits of heterogeneity or complementarity ("What I do not have, you do—and vice versa.")

Developing effective social networks

In addition to help from founding team members, entrepreneurs must also frequently obtain help and advice from other persons outside the team. As noted in Chapter 5, social networks are social structures composed of a set of members (individuals, organizations) and the ties between them, which indicate who knows whom, how well, and in what context(s). Large, broad, and effective social networks can provide entrepreneurs with helpful resources ranging from technical advice, financial resources, or referrals concerning the best person of company to use for a specific job, through psychological and emotional support. Overall, the value of a strong social network can hardly be overemphasized. In the world of business, as in many other contexts, who you know is often as important—or even more important—than what you know. And always remember these two points: (1) the more people you know and with whom you have established relationships, the more likely you are to find the help you need when you need it, and (2) people who do not know you will probably not think of you when they are seeking the kind of products, services, or ideas you and your company can provide. Where social networks are concerned, the effort required to build them (which can involve attending conferences, trade fairs, and many other activities) is usually well worthwhile.

Social and political skills: beyond social capital

In Chapter 5, we noted that social capital (which reflects individual's social networks, reputation, and experience)—gets them "through the door" (e.g., wins them the job interview, provides access to potential customers, etc.), but that social and political skills determine what happens once they are inside. That is one strong reason why social and political skills are so important for entrepreneurs. Why, for instance, should a potential customer order a new product from a relatively unknown company? Unless the entrepreneur can make a compelling case for this action (i.e., demonstrate a high level of persuasiveness), can make a good impression on the possible customer, and be perceived as sincere—someone the customer can trust—why should the customer take a chance? In fact, the odds of this happening are very small; the customer would prefer to stick with established companies and sources. This is where social and political skills enter the picture: they can often tip the balance in favor of the entrepreneur in many contexts. In business plan competitions (see Chapter 6), the prizes often go to the teams that make the most convincing presentations. True, other factors are important, too. But even so, entrepreneurs who can deliver their message very effectively often gain a significant advantage.

In short, for building social networks, for obtaining the support they need, and even for attracting first-rate employees, entrepreneurs need, and can benefit greatly from, excellent social and political skills. These are certainly valuable in almost any walk of life, but for entrepreneurs—who are, after all, attempting to change the world in certain ways and must often rely on their own skills rather than the established reputation of a large organization—they can be especially crucial. Fortunately, these skills can be acquired and strengthened—many techniques for accomplishing this task have been developed so even entrepreneurs who are not especially likable or persuasive, can attain significant improvement through personal effort.

Important steps along the way

Once entrepreneurs begin the process of converting their ideas into something real, they must perform many different tasks—and do so well—if they hope to succeed. (Remember, the odds appear to be heavily against them.) Among these tasks, though, three seem worthy of special mention: managing their own enthusiasm and passion, striking a good balance between planning and flexibility, and making good decisions.

Savor—but also manage—your enthusiasm and passion

As a group, entrepreneurs are amazingly upbeat; and that is hardly surprising, because optimism and a high level of enthusiasm (positive affect is a general term for such positive feelings) seem to go hand-in-hand with trying to create something new and better. In general, the tendency to be positive is beneficial. Similarly, entrepreneurs' passion for their activities, which involves an intense desire to act entrepreneurially, is also mainly a "plus". But as noted at many points in this book, there are limits to everything—a point beyond which there can really be too much of a good thing. In fact, as noted in Chapter 6, there is a major downside to being too positive too much of the time, and to feeling too strongly committed to entrepreneurial actions (e.g., being a founder or developer of a new venture; Murnieks et al., in press, 2012). For instance, persons very high in positive affect may be relatively unwilling to pay attention to negative input or information—a potentially devastating problem for entrepreneurs who definitely need all the feedback (especially negative) they can get. A key skill for entrepreneurs, then, is that of holding such tendencies in check so that they reap the benefits they provide, but incur fewer of the costs. The bottom line, then, is: by all means, be excited and upbeat, but hold these tendencies at least somewhat in check.

Plan—but remain flexible

Another essential skill for entrepreneurs also involves striking a balance between planning, which is essential for reaching important goals, and flexibility—the capacity to improvise, change direction, and use what is available instead of necessarily sticking to a pre-arranged plan. No, entrepreneurs are unlikely to attain financial and other resources without a detailed plan—perhaps even a formal business plan. But such plans should be viewed primarily as *guidelines*, rather than diagrams or maps for reaching key goals that must be carefully and fully followed. In fact, improvisation is often necessary; the strategy of effectuation—using what you have to determine where you want to go—can sometimes be helpful. Once again, therefore, the key may lie in balance between the pools of too much planning and too little.

Making good decisions—alone, and in groups

Yet another essential tool for entrepreneurs involves the capacity to make good decisions, on the fly and in the absence of complete information. When making decisions alone, individuals can utilize sharply contrasting approaches (see Chapter 7). On the one hand, they can attempt to follow a completely rational-analytic approach, in which all options are carefully examined, and the most suitable or attractive are carefully chosen. That can be very time-consuming, and often, requires more information than anyone working in uncertain and unpredictable situations has or can afford. Similarly, the strategy of *maximizing*—seeking the best possible solution or strategy is often not feasible for entrepreneurs. The information needed for such choices is not unavailable—and perhaps unobtainable because it does not exist—and the time required for making the "perfect" decision is excessive. Much better, then, is a strategy based on *satisficing*—seeking the first solution that works. This is what entrepreneurs generally do—especially successful ones—and if often yields decisions that are not only good and made quickly, but also leaves entrepreneurs with fewer doubts about their choices and fewer regrets about options they did not select.

Decisions made in groups pose additional dilemmas. Contrary to common sense, groups often make decisions that are more, not less, extreme than individuals. In addition, they often tend to focus on information most or all members have, while ignoring information held by only one a few members. Another potentially harmful tendency involves *groupthink*—the tendency for homogeneous decision-making groups to assume that they cannot make an error, and to "lock out" input from non-members. The fact that a decision is made in a group, therefore, is no guarantee that it will be a good one, and it is crucial for entrepreneurs to recognize this fact so that they take precautions against placing too much trust in such decisions.

Resilience in the face of the inevitable setbacks

The task of creating or implementing something new is a difficult one, and since entrepreneurs are, by definition, breaking new ground, they often face high levels of stress and are almost certain to make many errors. Stress can interfere significantly with performance, personal health, and if prolonged and intense, can lead to burn-out—a very damaging syndrome in which individuals lose all their enthusiasm for their careers or jobs (e.g., Berglas, 2001). So, being able to manage it effectively should be included on any list of key skills or tools for entrepreneurs. In addition, entrepreneurs need two further skills: the capacity to learn from their mistakes, and developing high levels of subjective well-being (personal happiness).

Learning from mistakes

Errors or mistakes, although painful, can also be a source of significant learning—and better performance the next time around. There are many techniques for maximizing the benefits of errors, but two are especially effective: (1) focusing on what might have been different, and so, perhaps, have generated better outcomes, and (2) learning to invite, and accept, negative feedback from others. The first involves *counterfactual thinking*—imagining what might have been, and when such thought focuses on why errors occurred and how they can be prevented in the future, it can be very beneficial. If instead, it focuses only on visions or more positive outcomes, it can result in strong feelings of regret and other negative reactions that are much less likely to facilitate learning.

Learning to accept criticism is another important skill for entrepreneurs. If, for instance, a potential customer decides against placing an order, it is important to learn why this happened, and to accept negative input from the customer not only graciously, but with an open mind. Doing so can facilitate learning from errors so that in fact, they are not repeated. Perhaps you have heard the satiric statement: "Experience is a wonderful thing; it helps me to recognize when I'm making the same mistake again." That's precisely the wrong view of the benefits of experience, which should, instead, contribute to learning, and therefore, to avoiding rather than repeating the same errors.

Personal happiness: the ultimate goal

Although entrepreneurs have many different motives and goals in mind when they set out to change the world, ultimately, whether these goals involve obtaining personal wealth, fame, increased independence, more interesting and fulfilling work, or anything else, the overarching goal should almost certainly be that of increased personal happiness. Entrepreneurs seeking wealth may find, to their surprise and disappointment, that it does not yield the increased happiness (subjective well-being) that they seek (e.g., Diener et al., 2010). But those seeking other goals—especially ones that fit under the general heading of "self-fulfillment"—may in fact gain in personal happiness as they move toward these more specific goals. In any case, it is important for entrepreneurs to keep this fundamental goal in mind as they move from ideas, to plans, to overt actions, and to success or failure. Overall, of course, there is no single formula or set of guidelines for attaining happiness. But given the powerful human urge to create, and satisfactions that can derive from merely trying to do so, entrepreneurship, in its many different forms, does seem to offer a unique—and uniquely fulfilling—route to this goal. So, my personal best wishes for your own entrepreneurial journey, whatever form it may take, wherever it leads you, and however it may unfold.

Summary of key points

Even though the odds are strongly against entrepreneurs, they can succeed if they build the tools—the skills, motives, capacities, and strategies—that are crucial for creating something new and better, and for making the *possible* real. The list of such skills is long (see Table 9.1), but all can be acquired or developed, and all have been identified by careful, systematic

Table 9.1 *Tools of highly effective entrepreneurs: essential equipment for making the possible real*

Tool or Factor	*Description*
Motivation	Understanding what you seek
Creativity	Breaking free of "mental ruts"
Identifying feasible opportunities	Actively searching, finding, and evaluating opportunities
Self-regulation	Self-control, knowing what you know and do not, and knowing when to "get out"
Building a strong founding team	Emphasize complementarity rather than similarity ("What I have you do not, and vice versa)
Developing effective social networks	Ones that are strong, broad, and can provide important resources
Social/political skills	Being able to get along well with others and form effective relationships with them
Manage your emotions and passion	Be positive, but keep within bounds!
Balance planning and flexibility	Plan carefully, but leave lots of room for improvisation and effectuation
Decision-making	Satisfice, do not maximize; and beware of groupthink, group polarization, and ignoring unshared information
Manage stress	Adopt effective (i.e., problem-focused) strategies
Learn from mistakes	Use counterfactual thinking well; learn to accept and use negative feedback
Build subjective well-being	Remember that personal happiness is the ultimate goal, and that wealth does not necessarily provide it!

Note: Although no complete list of the ingredients of entrepreneurial excellence may be possible (there are too many factors for this to be possible!) the ones shown here have been found, by careful research, to be among the most important—and the ones you should seek to develop.

research. In sum, in answer to the question "What *is* a highly effective or capable entrepreneur?" one useful answer seems to be: individuals who start this difficult journey with the necessary personal equipment. Once again, it is important to note that even if all the personal skills and capabilities listed in Table 9.1 are in place, that—in itself—is no guarantee of success. External conditions and factors beyond entrepreneurs' control (e.g., economic factors, political events, sudden shifts in public preferences, the actions of competitors) can, and do, play important roles. But, such factors aside, entrepreneurs do play a key role in the process, so it is always useful for them to keep these words, offered by Margaret Thatcher, the well-known British Prime Minister, firmly in mind: "People think that at the top there isn't much room. They tend to think of it as an Everest. My message is that there is tons of room at the top." In short, you can get there, but it takes not just energy and persistence, but careful preparation too!

References

Baron, R.A. (2012). *Entrepreneurship: An evidence-based guide*. Cheltenham, UK, and Northampton, MA, USA: Edward Elgar Publishing.

Berglas, S. (2001). *Reclaiming the Fire: How successful people overcome burnout*. New York: Random House.

Diener, E., Ng, W., Harter, J., and Arora, R. (2010). Wealth and happiness across the world: Material prosperity predicts life evaluation, whereas psychosocial prosperity predicts positive feelings. *Journal of Personality and Social Psychology*, 99, 52–61.

Murnieks, C.Y., Mosakowski, E., and Cardon, M.S. (in press, 2012). Pathways of passion: Identity centrality, passion, and behavior among entrepreneurs. *Journal of Management*.

Sternberg, R.J., and Sternberg, K. (2011). *Cognitive Psychology*. Cincinnati: Cengage.

Author index

Aarstad, J. 90
Adler, P. 90
Ajzen, I. 27
Aldrich, H.E. 87, 90
Alquist, J.L. 66–7
Alvarez, S.A. 55
Amabile, T.M. 50
Amodio, D.M. 41
Argyris, C. 158–9
Ariely, D. 41, 43, 47, 106
Ashby, F.G. 111
Avey, J.B. 14, 153

Baas, M. 111
Baker, T. 120
Bandura, A. 25, 27–8, 104, 118
Barney, J.B. 55, 106, 140
Baron, Robert 14–15, 30–31, 33, 40, 42, 44, 46, 48–9, 55–6, 58, 64, 67, 71–2, 76, 95–6, 105, 107, 109–11, 113, 122, 133, 150, 153–4, 156, 158, 164, 173
Barrick, M.R. 108
Barringer, B. 71
Barsade, S.G. 110
Baum, J.R. 34
Baumeister, R.D. 9, 65–7
Beach, L.R. 131
Beehr, T. 150
Bell, A.G. 6–7
Belliveau, M.A. 93
Berglas, S. 151, 182
Bledo, R. 83
Bledow, R. 111
Bonaccio, S. 87
Boyce, D.J. 163
Branigan, C.A. 56
Branscombe, N.R. 48, 109–10
Brazeal, D.H. 140
Bretz, R.D. 10
Brinckmann, J. 117
Burt, R. 91
Busenitz, L.W. 106, 140
Bush, G.H.W. 133
Bush, G.W. 142
Buss, D.H. 108
Byrne, Donn 85

Camus, A. 129, 132
Cardon, M.S. 68, 97, 111, 114, 160
Cassar, G. 30
Cavendish, M. 12
Chandler, G.N. 28
Chandler, J.J. 44
Chan, M. 153
Chao, R. 151
Chen, X.P. 97, 115
Ciavarella, M.A. 109
Clinton, B. 133
Cohen, S. 153
Collins, C.J. 28
Cope, J. 161
Corbett, A.C. 118
Costa Jr., P.T. 27
Cote, S. 71
Cropanzano, R. 111
Cyders, M.A. 72

Dalal, R.S. 87
DeNisi, A.S. 92–3
De Young, C.G. 112
Denson, T.F. 67
Dew, N. 122
Diener, E. 20, 93, 153, 162–3, 183
Domjan, M. 155
Downs, A.C. 93
Drucker, P. 120
Duckworth, A.L. 68
Dunegan, K.J. 131
Dunlosky, J. 76
Dunning, D. 159

Einstein, A. 113
Ellsworth, P.C. 47
Ensley, M.D. 58

Ferris, G.R. 15, 91–2
Figner, B. 105–6
Fine, S. 10
Flavell, J. 75–6
Folkman, S. 151
Ford, J.K. 75

Forgas, J.P. 64, 111
Fredrickson, B.L. 56, 111, 113
Fried, Y. 31
Fujita, K. 67

Gaglio, C.M. 55
Gates, B. 3, 7–8, 12, 19
Gawande, A. 157
George, J.M. 110–11
Ghoshal, S. 90
Gibson, D.E. 110
Gigone, E. 143
Gladwell, M. 7–8
Goleman, D. 70
Granovetter, M.S. 91
Grant, A. 105, 112–13
Greenberg, J. 133
Gregoire, D.A. 55, 59
Grewal, D. 70–71
Griffitt, W. 85
Gupta, N. 53

Hardman, D. 128
Hastie, R. 143
Haynie, J.M. 76
Hayward, M.A.L. 161
Henry, R.A. 42
Hertzog, C. 76
Hirt, E.R. 143
Hmieleski, K.M. 46, 105, 118
Hochwarter, W.A. 93

Ilies, R. 111
Ireland, D. 71
Isen, A.M. 111
Ivancevich, J. 150

Janicki-Deverts, D. 153
Jansen, E. 28
Jarvstad, A. 134–5
Jex, S. 150
Jobs, Steve 3, 19, 142
Johnson, L.B. 142
Judge, T.A. 111
Jullerat, T. 31

Kahneman, D. 135
Kaplan, S. 111
Katz, J. 55
Khaire, M. 112
Kiecolt-Glaser, J.K. 150
Kim, P.H. 87, 90
Kirzner, I. 55–6
Knight, A.P. 110
Komisar, R. 118
Kotsou, I. 92, 96
Koutstaal, W. 12
Kray, L.J. 156
Krieger, M. 3
Krueger N.J. 140
Kruglanski, A.W. 72
Kurtz, M.M. 96
Kwon, S. 90

Lange, J.K.E. 117
Larsen, J.T. 163
Larson, R. 108
Latham, G.P. 24
Lazarus, R.A. 151
Lehrer, J. 75
Lewicki, R.J. 93
Locke, E.A. 24, 34
Lopes, P.N. 70–71
Losada, M.F. 113
Lux, S. 97
Lyons, P.M. 93
Lyubomirsky, S. 111, 164

Markman, G.D. 71, 95, 104, 143
Matlin, M.W. 56
Matteson, M. 150
McCaffery, T. 53–4
McCrae, R.R. 27
McKibban, A.R. 163
McMullen, J.S. 50, 55, 78
Medvec, V.H. 156
Melton, R.J. 111
Meucci, Antonio 6–7
Miner, J.B. 44
Miner, J.V. 105
Mischel, W. 73–5
Mitchell, T.R. 131
Mojzisch, A. 141
Mount, M.K. 108–9
Mueller, J.S. 50
Mueser, K.T. 96
Mullins, J. 118
Murnieks, C.Y. 111, 114–15, 149, 181

Nahapiet, J. 90
Nambisan, S. 67, 70, 77
Nelson, R.A. 120
Nietzsche, F. 148–9
Nisbett, R.E. 12
Nocera, Prof D. 31–2, 138

Obama, B. 133
O'Brien, E. 47
Oishi, S. 112–13, 163
Owen, D. 32

Page, L. 12
Peterson, S.J. 14, 153
Ployhart, R.E. 10
Portes, A. 90
Pratkanis, A.R. 86
Pronin, E. 44
Putnam, F. 90

Quinn, P.D. 68
Quoidbach, J. 163

Raju, N.S. 44, 105
Ramsey, D. 73
de Ridder, D. 66–7
Riggio, R.E. 92, 95
Rindova, V. 29–30
Roberts, R.E. 164
Robbins, T.L. 92–3
Roese, N.J. 155
Roth, P.L. 105, 140
Rowling, J.K. 47
Rueff, M. 85, 87

de Saint-Expurey, A. 117
Salovey, P. 70
Sanna, L.J. 156
Sarasvathy, S.D. 122
Schaich, Ron 19
Schmidt, A.M. 75
Schneider, B. 154
Schulz-Hardt, S. 141
Schwartz, B. 105, 112–13, 136–8
Seibert, S.E. 109
Seligman, M.E.P. 93
Shane, S.A. 3, 33, 49
Sharot, T. 46, 47
Shepherd, D.A. 50, 55, 59, 76, 78, 161–2
Sheppes, G. 72
Sternberg, K. 41, 54, 174
Sternberg, R.J. 41, 54, 174
Stewart, W.H. 105, 140
Struck, F. 53
Systrom, Kevin 3

Tang, J. 15, 50, 55–6, 96
Throckmorton, B. 92
Tice, D.M. 72
Tierney, J. 65, 67
Topolinski, S. 53
Trump, D. 104
Tung, R.L. 33
Turner, M.E. 86
Twain, M. 66
Twitter 39

Ucbasaran, D. 160–61

Vallerand, R.J. 114
Vohs, K.D. 9

Ward, T.B. 50
Washington, G. 155
Wayne, S.J. 93
Weber, E.U. 105
Weick, C. 162
Weiss, H.M. 111
Wilson, T.D. 12
Wright, M. 164

Xie, J.L. 150
Xu, J. 164

Zahra, S.A. 96
Zelaznock, T. 13
Zhao, H. 27–8, 104, 109, 111
Zuckerberg, M. 2–3, 12

Subject index

affect infusion
concept of 46
alertness
components of, 55–7
attraction-selection-attrition (ASA) theory 103, 106
application to entrepreneurship 10
concept of 10

bias processing of information
in decision making 131, 143–6
big five dimensions of personality 27–8, 108–10
bricolage
concept of 120
business failure 161, 165
financial costs of 160, 165
psychological costs of 160–61, 165
social costs of 160, 165
strategies to cope with 161–2

cognitive traps
in decision-making 132, 134
complementarily
concept of 87
concepts 49, 54, 59
role in creativity 50–52, 174
confirmation bias
concept of 132–3
impact on entrepreneurship 44
counterfactual thinking 182
affects of 156
concept of 155–6
criticism 73, 159
constructive 158
destructive 158
giving 157, 165
importance of 157–8
receiving 157–9, 165, 183

decision-making 128–9, 134, 139, 144
as cognitive process 132
biased processing of information 131, 143–6
cognitive traps 132, 134
effective 135
group 141–2, 144
group polarization 141–2, 146
impact of subjective well-being on 136
maximizing 136, 144, 181
person sensitivity bias 144
rational analytic model 129–31

effectuation
concept of 119–21, 123
opposing relationship to business plan 121–2
emotional contagion 110
emotional intelligence (EI)
concept of 70–71
use of 71
entrepreneurs 16, 20, 64–5, 67, 84, 99, 103, 114, 118–20, 140, 171, 173–4
application of ASA to 10
as process 49, 59
business plan 116–17
cognitive processes in 39–43, 138
definition of 1–4, 9–10
development of 58, 97, 183
emotional stability of 110–11
founding teams of 84–7
guidelines 181
impact of confirmation bias on 44
impact of optimistic bias on 46, 48
importance for delay of gratification 74–5
importance of improvisation 118–19, 122–3, 136, 153, 181
intentions of 20, 22, 26–9, 35, 104
media depiction of 2–3
motivation of 22, 26–35
passion 109–10, 113–16, 123
pursuit of new ventures 174–5
role in failure of ventures 6
role of tools in success of 11, 16–17
self-efficacy of 27–8
stress levels experienced by 13–16, 154
use of EI in 71–2
use of social networks in 179

use of subjective criteria by 140
view as risk-takers 105–7, 122
entrepreneurial intentions 20, 22, 26–9, 35, 104
escalation of commitment
concept of 134
expectancy theory
concept of 23–4
factors involved in 23

failure
business 2, 5–6, 28, 71, 149, 159–62, 165
funding 47
sources of 28

goal-setting theory
concept of 24–5
group decision-making 141–2, 144
group polarization 141–2, 146
groupthink 99, 182
concept of 86

heuristics 64
anchor-and-adjustment 45
concept of 45
higher mental processes 110

image theory
concept of 131
implicit favourite
concept of 132
improvisation 118–19, 122–3, 136, 153, 181

learning 58, 75, 113, 123, 155, 161
from mistakes 155–7, 160, 162, 165, 182–3
locomotion
concept of 72

maximizing 181–2
versus satisficing 136–9, 144
memory 41–2
auto-biographical 42
fallibility of 12–13, 58–9
long-term 42
procedural 42–3
mental ruts 51–2, 174
motivation
of entrepreneurs 22, 26–35
psychological concept of 21–2, 25

opportunity costs
concept of 138
opportunity recognition 54–7, 59
optimistic bias 46, 48, 59
planning fallacy 46
role of pattern recognition 58, 175

passion 109–10, 113–16, 123
pattern recognition 56, 58, 91, 175
concept of 59, 138
perceived behavioural control
concept of 27
person sensitivity error
concept of 133
personality aspects 109, 122–3
agreeableness 27–8, 108
conscientiousness 27, 108
emotional stability 27, 108–10
extraversion 27, 108
openness to experience 27, 108
planned behaviour theory
concept of 26
political skills 94–5, 97–9, 179–80
forms of 93
impact on development of social networks 94
role in organizational processes 92–3
strengthening of 96–8
psychological capital
concept of 14
levels of 153–4

rational analytic model of decision making 129–31
reasoned action theory
concept of 26
Remote Associates Test
concept of 53
related research 53–4
risk 13, 28, 44, 72, 107, 129, 171
acceptance of high levels of 105–6
management of 122, 140

satisficing 182
versus maximizing 136–9, 144
self-efficacy 28, 153, 161
concept of 104–5, 115, 122
enhancement of 115
of entrepreneurs 27–8
self-fulfillment 26, 183
desire for 32–3
self-knowledge 15, 63
importance of 48, 177
self-regulation 64–5, 113, 176–7
concept of 63, 65, 78
delaying gratification 63–4, 73–5
emotional 72

GRIT 68–70, 79
metacognition 75–9, 177
self-control 66–7, 70, 78–9, 176–7
signal-detection theory 145
concept of 139–41
examples of 140
similarity (homophily) 87, 99
social capital 98, 179
concept of 90
social ties 90–91
use of 95
social network theory 91
focus of 93
social networks 91–2, 108–9
concept of 89, 99, 179
development of 93–4, 98, 180
network composition 94
network effectiveness 94
network efficiency of 94
use of 88–9, 179
social skills 92, 95–6, 98–9, 179–80
forms of 93
strengthening of 96–8
impact on development of social networks 94
stress 149–50, 153
levels experienced by entrepreneurs 13–15
relation to physical health 150–51, 165
strategies for management of 151–2
subjective well-being 164–5, 182
impact on decision-making 136
lack of impact of wealth on 162–3, 165–6
success
factors responsible for 12–13
measurements of 62–3
role of passion in, 113–14
sunk costs 134, 177
concept of 46–7

tools 1–2, 8, 11, 33, 51, 65, 129, 171
examples of 15–16
potential lack of 7, 171
role in success of entrepreneurship 7, 11, 16–17, 40, 63, 68, 149, 159–60, 171, 182–3

venture capitalists (VC) 28, 47, 95
development of social networks by 89–90
financial backing provided by 22, 34, 115–17, 133–4, 178
potential guidance role of 64
ventures 24, 56, 58, 75–6, 83, 105–6, 131, 140, 149, 176, 179
average lifespan of 4–6
creation of 16, 20, 22, 27–8, 30, 40, 49–50, 68, 86, 98, 104
failure of 4–6, 134, 159–61, 165, 170–71
financial earnings from 75, 95, 113
founding teams/founders of 83–5, 89, 110, 117, 178–9, 181
funding of 33, 47, 133
impact of grit in performance of 70
impact of entrepreneurial passion on 123
impact of self-control in performance of 67
potentially limited resources of 142
role of guidance system in 34